Ferdinand Freiherr von Mueller, Ferdinand von Mueller

The vegetation of the Chatham Islands

Ferdinand Freiherr von Mueller, Ferdinand von Mueller

The vegetation of the Chatham Islands

ISBN/EAN: 9783337374303

Printed in Europe, USA, Canada, Australia, Japan

Cover: Foto ©berggeist007 / pixelio.de

More available books at **www.hansebooks.com**

THE

VEGETATION

OF THE

ĊHATHAM-ISLANDS,

SKETCHED BY

FERDINAND MUELLER,

PH.D., M.D., F.R.S., F.L.S., F.R.G.S., C.M.Z.S.

'BELEHRUNG FORDR' ICH, WAHRHEIT UND ERKENNTNISS."—*Chamisso.*

By Authority:
JOHN FERRES, GOVERNMENT PRINTER, MELBOURNE.
MDCCCLXIV. &c.

TO

WILLIAM THOMAS LOCKE TRAVERS, ESQ., F.L.S.,

BARRISTER-AT-LAW,

FORMERLY JUDGE OF THE SUPREME COURT OF THE PROVINCE OF CANTERBURY, NEW ZEALAND,

THROUGH WHOSE

DISINTERESTED GENEROSITY AND HIGH-MINDED ZEAL FOR SCIENTIFIC RESEARCH
THE PHYTOLOGICAL EXPLORATION OF THE CHATHAM-ISLANDS
WAS MAINLY ACCOMPLISHED,

AND BY WHOSE

PERSONAL EXERTIONS OUR KNOWLEDGE OF THE VEGETATION OF NEW ZEALAND
HAS BEEN SO EXTENSIVELY ADVANCED,

This small Volume is Dedicated,

WITH THE AUTHOR'S MOST GRATEFUL REGARDS.

INTRODUCTION.

THIS world, the abode of the human race, is so wonderfully replete with organic life, and the harmonious perfection of each organism reveals so gloriously the divine power of the Supreme who created the universe, that the elucidation of His living works in nature around us must ever be one of the most elevating designs of the intellect of man. The mind longs to reflect on the destiny assigned to each in the great economy of nature, and to shed light as far as it is within wordly power on all as yet obscure. Hence our desire to penetrate every locality occupied by living creation; hence the persuasion that nothing is too insignificant to be drawn into the sphere of an enlightened enquiry. Not until the searching naturalist has traversed every spot of the globe can he fully understand and appreciate the relation and affinities of its various natural productions; not until then can a complete and logical arrangement of the animals and plants contemporaneous with the human species be accomplished.

To offer a trifling contribution towards such an object is the aim of these pages. If under any circumstances a deep interest is attached to isolated islands, which harbour perhaps the only remnant of the vegetation or the animal life of countries ages ago sunk beneath the ocean, this interest cannot be otherwise than most vividly excited in regard to the "Chatham-Islands," inasmuch as this little group is the last eastward of New Zealand, no further land existing under those latitudes in the wide interjacent oceanic space until the west-coast of South America is reached.

Before entering into a detailed exposition of the plants of those islands, which is the main purpose of this writing, it may not be inappropriate to review the results of previous exertions for the revelation of their flora.

A

Dr. Joseph Hooker, the accomplished author of the large work on the plants of New Zealand, recognizing the importance of an investigation of the Chatham-Islands, was anxious whilst attached to Sir James Ross's antarctic expedition to visit this group; but the "Erebus" and "Terror" when near the spot encountered such foggy weather as to prevent the officers from sighting the coast and oblige them to abandon their plan of landing. But in the previous year (1840) it fell to the share of Dr. Ernst Dieffenbach, the German translator of Charles Darwin's excellent work on his travels, to explore these islands on behalf of the New Zealand Company; and although mainly engaged in geographic, geognostic and agronomic researches, of which a lucid account is given in the Journal of the Royal Geographical Society of London (vol. 11, pp. 195–215), this well-informed natural philosopher did not fail to render simultaneously known some of the main features of the vegetation, and brought the first plants from these islands to England; the records of these have been inserted in Dr. Hooker's work on New Zealand plants. How far the examination of the plants collected during the voyage of the frigate "Venus" has thrown light on the vegetation of the Chatham-Islands, the writer is unable here to ascertain, not having had access to the important publications which emanated from that expedition. However, amongst the Chatham plants, illustrated by Monsieur De Caisne in that work, solely Eurybia semidentata is quoted by Dr. Hooker. In 1858 a temporary direct trade between Melbourne and the Chatham-Islands brought within the reach of the author several plants not embraced in Dr. Dieffenbach's collection; these were kindly gathered and presented by Captain Anderson; others were received about the same period through the friendly offices of the Honorable Dr. Featherstone, Administrator of the Government in the Province of Wellington, and in the same year the writer was favoured by Mr. A. J. Ralston, at the time a citizen of Melbourne, with a gift of flower- and fruit-bearing plants of the magnificent Myosotidium nobile. Shortly afterwards an essay on that singular plant, described then as Cynoglossum Chathamicum, was read before the Philosophical Institute of Victoria, but was withdrawn from publication in the Transactions of that Society, because meanwhile the venerable Sir Will. Hooker had given an account of the same plant; moreover, it received simultaneously from Dr. Hooker the name

Cynoglossum nobile in Professor Lindley's Gardeners' Chronicle. Three years afterwards Mr. Will. Seed, of Wellington, proceeded to the Chatham-Islands on behalf of the Customs Department of New Zealand. The published official Report resulting from this gentleman's journey (bearing date 21 October, 1861) contains also some interesting ethnological information and many intelligent remarks on the physical aspect, on the resources and on the natural productions of the islands, and is accompanied by a list of the trees observed there. Valuable as all this information proved for phytology, it suggested only more strongly the desirability of effecting a methodical search of the plants of this group. Therefore it must be considered a singularly fortunate circumstance, that the gentleman to whom this work is dedicated, on learning through Dr. Julius Haast, the distinguished geographer and geologist of Canterbury, how advantageous to science a close examination of the Chatham-Islands might prove, at once decided to entrust this task to his son, Mr. Henry H. Travers, and with a rare generosity Mr. Travers resolved to bear solely the whole of the expenses arising from this mission. If such acts of munificence were more frequent how delightfully and speedily might many scientific enquiries, for which the opportunities are too often lost, be originated or sustained. Thus the youthful naturalist set out on his voyage in October, 1863, and during a stay of several months on Chatham- and Pitt-Islands accumulated the material, which forms the principal foundation of this memoir. Moreover, the young gentleman added considerably to our knowledge of the physical geography and zoology of these insular spots. These observations and his anthropological notes will elsewhere receive due promulgation. But it is only just to offer on this occasion a public acknowledgement of the great kindness experienced by the traveller from Mr. Frederick Hunt, of Pitt-Island, who not only afforded him hospitality and aid in that place, but also periodically joined in his excursions through Chatham-Island. Captain Thomas, Revenue Officer, and Mr. Alexander Shand, are likewise entitled to recognition of services rendered to Mr. Travers whilst forming his collections.

The fullest set of the plants secured was after Mr. Travers's return presented by his father to the Phytological Museum of Melbourne with a view to their elucidation.

The pages of the first volume of Dr. Hooker's Manual on New Zealand plants having passed through the printing-press before the Chatham-collections became available, the results of the present treatise could not be incorporated into that work, but will be accessible for any appendix to the second volume. The author regrets not having had the advantage of consulting Dr. Hooker's publication, which has not yet reached Australia; but it may be anticipated, that the recent observations of that illustrious phytographer on the New Zealand plants and those on the Chatham-Flora recorded in this book, have led in many instances to the same conclusions.

The collections thus accumulated in the Chatham-Islands comprise 129 species of plants, apparently indigenous. Of these 42 are dicotyledonous, representing not less than 32 orders and 37 genera; 20 are monocotyledonous (including a doubtful Calorophus mentioned by Dr. Hooker), and these exhibit 9 orders and 19 genera; the remaining 67 plants are acotyledonous. It may however be predicted, that although we are now fully acquainted with the trees and shrubs of these islands, still some phanerogamic herbaceous plants, most likely also several of the ferns common in New Zealand, and certainly an abundance of cryptogamic plants remain still to be discovered. Plants evidently immigrated are however not taken into account; for instance, Cerastium vulgatum, Potentilla anserina, Taraxacum officinale, Sonchus oleraceus, Solanum nigrum, Holcus lanatus. In addition to these introduced plants Mr. Travers mentions in his journal the English Daisy, the Mustard, the Dock, the Burr, a Polygonum (probably P. aviculare) and the Wild Strawberry. Of ferns, inclusive of Lycopodiaceæ, we know now 25 as existing in the Chatham-Islands, all identical with species found in New Zealand; these comprise 14 genera. The other cryptogamic plants collected by Mr. Travers, have not been systematically enumerated in this work, inasmuch as they evidently present an imperfect picture of that part of the Flora of this island-group. May it therefore suffice on this occasion to point out, that the 12 Mosses of the collection include representatives of the genera Sphagnum, Rhacomitrium, Macromitrium, Bryum, Funaria, Polytrichum, Hypnum, Hookeria. The only Lichenaster is devoid of fruit. The 9 Lichens are comprehended in the genera Ramalina, Peltigera, Sticta, Parmelia, Stereocaulon and Cladonia; the 20 Algæ comprise species of Sargassum,

Cystophora, Phyllospora, Ecklonia, Sphacelaria, Champia, Amphiroa, Melobesia, Plocamium, Iridæa, Ballia, Griffithsia, Caulerpa, Conferva (Chætomorpha) and some other genera. Of Fungi none are contained in the collection. A comparison of the plants of the Chatham-Islands with those of New Zealand, as far as they are known to the author, has proved, that only 9 phanerogamous species are peculiar to the former, and even of the specific validity of some of these we are not finally assured. Eight of these seemingly endemic plants are referable to the genera Coprosma, Gingidium, Eurybia, Senecio, Leptinella and Myrsine, whilst only one (Myosotidium nobile) constitutes a genus peculiar to the islands. As all the other genera are represented in the vegetation of New Zealand, we are warranted in regarding the flora as merely an extension of that of the great neighbouring group, the territory of which may indeed at one period have stretched uninterruptedly across, a supposition strengthened by the fact of navigators having obtained soundings throughout the interjacent sea. This persuasion impresses itself the more forcibly on the mind of the observer, if he contrasts the Chatham-vegetation with that of Raoul-Island, imperfectly as it is known (conf. J. Hooker in Proceed. Linn. Soc. i. 125) and more particularly with that of Norfolk-Island and of Tasmania, certain points of all these isolated spots being nearly equidistant from New Zealand. Little in the vegetation of the Chatham-Islands is strikingly peculiar, except the Myosotidium and the grand arborescent Compositæ (Eurybia Traversii and Senecio Huntii), which add two singular forms to the very limited number of such plants of this extensive order as rise to the height of trees ; such also elsewhere occur chiefly in insular localities. Of the phanerogamic plants of the Chatham-Islands 24, and of the ferns 6, are otherwise restricted to New Zealand. Other 11 of the di- or monocotyledonous plants and seemingly 6 of the ferns are common to New Zealand and Australia or some of the adjacent islands, whilst 12 of the phanerogamic plants (the introduced ones excluded) and as many of the ferns are more or less widely dispersed over the globe. Of the Australian plants, occurring in the Chatham-group, only one, Leucopogon Richei, seems hitherto not observed in New Zealand ; whereas Hymenanthera latifolia, indigenous both to Norfolk- and Chatham-Island, represents in all likelihood but one of the extreme forms of an Australian and New Zealandian species. Tetragonia implexicoma, frequent on the shores of extratropical Australia and the Chatham-Islands, has also been recently discovered by Mr. Travers, sen., at Lyttleton-Harbour, of New Zealand, where apparently it produces its red baccate fruit more readily than elsewhere.

Dr. Dieffenbach alludes to the occurrence of two plants in the Chatham-Islands not otherwise as such on record, namely a Typha and Cyathea medullaris, which latter will probably be found identical with the Cyathea Cunninghami of this work.

The berry-bearing supposed pine, mentioned by Dr. Dieffenbach as a Taxus, is likely to be identified with Cyathodes acerosa.

Some of the notes of Mr. Travers's journal, as bearing directly on the vegetation of these islands, need here an abridged record.

Not only the Maori-huts, but also the dwellings of the Europeans are built of posts of fern-trees, lashed together with Supplejack (Rhipogonum scandens). The Toi-grass (Arundo conspicua) is employed for thatching. The aboriginal inhabitants, before their subjugation by the New Zealand natives, chiefly resorted to fern-roots as vegetable food prepared in the manner customary in New Zealand. They formed rafts of the flowerstalks of Phormium tenax, spliced with Supplejack, the generality of the timber not being sufficiently large for constructing canoes. It is therefore the author's intention to aid in introducing into the islands some of the Australian timber-trees of the most vigorous growth. The species of Potamogeton and other water-weeds, common in lakes of New Zealand and Australia, were not observed by Mr. Travers, who indeed in all his excursions noticed only one floating water-plant, to which however he could not obtain access. The peat in some localities was found to extend to a depth of fifty feet; in several parts of the island this peat has been on fire for years, burning at a considerable depth below the surface, which, when sufficiently undermined, caves in and is consumed. Mr. Travers saw ashes of these fires arising from a depth of thirty feet. At one place he remarked in the burning peat six or seven feet below the surface the trunks of trees, evidently far exceeding in size any now growing on the islands. Large numbers of indigenous herbaceous plants have probably been destroyed, partly by burning of the surface-ground and partly by depastured animals roaming over it.

The author feels, that it devolves on him to vindicate the principles, by which he was guided in the limitation of some of the species, circumscribed anew on this occasion; especially as the application of these principles involved the suppression of numerous plants, to which specific value had hitherto been unquestionably assigned. The description of a genuine species clearly should be so framed, as to admit of its embracing any of the aberrations from the more usual type, to which under various climatic or geologic circumstances a species can possibly be subject; and the diagnosis should be so constructed as to include all the cardinal characters of the species, none of these ever admitting of exceptions.

But the material even in the greatest museums and the evidence extant from the most extended field-observations are as yet insufficient for fixing finally the diagnosis of almost any known species; and it will be therefore a problem of future research, to trace anew the specific demarcation of the organic forms throughout the creation, to ascertain to what extent nature has endowed each to accommodate itself to altered conditions, to elucidate how far in each special instance such influences change the external form of the species, and to determine the bearing of each to the geological features of the globe. From his own lengthened observations, carried on in this sense, the author feels justified in drawing the conclusion, that the number of species has been vastly overrated, and further, that their distinction never rests on a solitary or on faint characters. A study of plants growing in localities, where they are exposed to most unusual agencies, yields results of profound significance; and the revelations, to be derived from a clear insight into the vegetation of Australia, are in many cases as startling as replete with deep instructive meaning. It is there where we may trace plants, often in strangely altered forms from the glacier-regions to forest depressions, in the mild air of which even tropical species may luxuriate; it is there again where we may witness the effects of the Sirocco of a desert-country on otherwise alpine or tropical jungle-plants.

But recognizing this wonderful adaptability of the species to sometimes singularly different circumstances, the writer has never been led to assume, that limitation of species is hopeless, or that an uninterrupted chain of graduations absolutely connects the forms of the living creation. Analytical dissections, counting by hundreds of thousands, instituted as well on living plants in the field as on the material accumulated in his museum, have never left such impressions on his mind ; but on the contrary convinced him of the great truth, that the Supreme power to which the universe owes its existence, called purposely forth those wonderful and specifically ever unalterable structures of symmetry and perfection, structures in which a transit to other species would destroy the beautiful harmony of their organization, and would annihilate their power to perform those functions specially allotted to each in this great world from the morn of creation to the end of this epoch. But be it understood, nature only created species, occasionally but not permanently obliterated in their characters by hybridism. Genera and orders are merely the strongholds around which we arbitrarily array them to facilitate generalization, to ease the search and to aid the memory. Hence the limitation of these must depend entirely on the individual view of the observer, and therefore be ever vacillating ; but this should not finally be the fate of the species.

The author, in conclusion, cannot suppress a hope of seeing our collections of animals and plants of the Chatham-group completed through the exertions of those residents who feel an interest in the works of nature which surround them. For scientific researches in lands mainly occupied by savages, the missionaries enjoy unparalleled facilities. But although these self-denying bearers of civilization have exercised throughout the last centuries a not unimportant influence on the task of our storing together the treasures of natural science, an influence never more gloriously exemplified than in the labours of Dr. Livingstone during our time, it cannot be concealed, that if the devoted men, who carry abroad the *word of God*, were more generally cognizant how often it is alone as yet in their power to reveal also many of the marvellous *works of God*, of which no spot, however desolate and lonely, is devoid, the universal history of nature would be much earlier written, and on the divine labours of the mission throughout the globe would be shed an additional brilliant lustre.

Botanic Gardens,
Melbourne, 15th September, 1864.

ENUMERATION

OF THE

PLANTS OF THE CHATHAM-ISLANDS.

VIOLACEÆ.

HYMENANTHERA LATIFOLIA.

Endl. Prodrom. Flor. Insul. Norfolk. 71 ; Endl. Iconograph. Gener. Plant. 108.

Varietas Chathamica ; leaves serrulate.

Rare in woods.

The specimina brought by Mr. Travers are fruit-bearing and respond fully to Endlicher's description and Bauer's illustration, with exception of the margin of the leaves, which is toothed by short regular notches. The stipules, overlooked in the plant of Norfolk-Island, and also by some describers of the Australian species, are akin to those of H. Banksii, although proportionately larger. The plant from Norfolk-Island and Chatham-Island may exhibit merely a luxurious variety of the Australian species. The embryo of the Chathamian plant agrees in its characters entirely with the united Australian and Tasmanian species ; the cotyledons being orbicular, not narrow as (probably from a side view of those of H. crassifolia in Fitch's Illustration, vii. of J. Hook. Fl. Nov. Zeal.) recorded in Benth. and J. Hook. admirable Genera Plantarum, i. p. 120, and Benth. Flor. Austr. i. 104. Having had only an opportunity of examining flowering specimens of the New Zealandian congener, which in no respect differ from certain especially alpine states of H. Banksii, the author must leave it for future observers to ascertain whether in reality H. crassifolia can be distinguished by the narrowness of its cotyledons. In adopting for the combined Australian and

B

Tasmanian plant the name H. Banksii, the former specific appellations H. angustifolia and H. dentata were abandoned as carrying no longer any significance with them, since it was moreover proved (Plants of Victoria, i. p. 70) that the form of the leaves is not even available for a precise distinction between H. Banksii and H. latifolia. In all probability only one species of this genus exists, for which the collective name H. Banksii might well be retained. R. Brown, when establishing the genus in his learned memoir on the Congo plants (Append. to Tuckey's Narrative of an Expedition to the River Zaire, p. 442), attributes to it the exceptional ordinal character of a two-celled berry ; all fruits, however, examined on this occasion, even in a very young state, proved one-celled.

LINEÆ.
LINUM MONOGYNUM.

G. Forster, Florul. Insular. Austral. Prodrom. 145; Botanic. Magaz. 3574; A. Richard, Voyage de l'Astrolabe, i. 317-318; Planchon in Hook. Lond. Journ. of Botany, vii. 170; J. Hook. Flor. Nov. Zeel. i. 28.

Almost everywhere except in woods.

GERANIACEÆ.
GERANIUM DISSECTUM.

Linné, Spec. Plant. 956; J. Hook. Flor. Nov. Zeel. i. 39; F. M. Plants of Victoria, i. 173; Benth. Flor. Austral. i. 295 ; G. pilosum, G. Forst. Prodr. 531; Sweet, Geran. ii. t. 119; A. Rich. Voy. de l'Astrolabe, i. 295.

Frequent except in woods.

MALVACEÆ.
PLAGIANTHUS BETULINUS.

All. Cunningham in Annals of Natural History, iv. 25; J. Hook. Flor. Nov. Zeeland. i. 29; P. urticinus, A. Cunn. l. c.

Chatham- and Pitt-Island ; common in woods. Native name " Whawhi."

The specimina are in fruit; these, the leaves and calyces are larger than those of the New Zealandian specimina of our museum.

From various parts of New Zealand our collection possesses plants of Plagianthus Lyallii (Asa Gray, Botany of Unit. States Explor. Expedition, 181), although only in a flowering state. These show the number of styles to be variable from 10 to 15 (10, 11, 13, 15), whilst the stigmata are oblique-terminal, very slightly or hardly decurrent, sometimes more curvate-clavate, sometimes so thickened as to appear capitate and to offer a transit to the more decidedly peltate stigma of Hoheria populnea. Thus the habit of the plant, its bisexual flowers and its stigmata mediate a clear transit to Sida, from which the closely allied Hoheria populnea, in appreciation of its winged carpels, can be also only sectionally separated, as indicated in the Fragm. Phyt. Austr. i. 29. A curious variety of Sida Lyallii, with small deeply incised leaves, collected by Mr. Travers and Dr. Haast in Middle Island, may be distinguished as *var. ribifolia*.

CORIARIEÆ.

CORIARIA RUSCIFOLIA.

Linné, Spec. Plant. 1467, accord. to J. Hook. Flor. Nov. Zeel. i. 45; C. sarmentosa, G. Forst. Florul. Insular. Austral. Prodr. 733; G. Forst. de Plant. Escul. Insul. Ocean. Austr. 46; Ach. Rich. Voy. de l'Astrolabe, i. 364; Bot. Magaz. 2470.

Common in woods and at their edges. Native name "Tutu."

The Chatham-plant represents the large-leaved normal form. The single orbicular or ovate-orbicular bracteoles at the base of the pedicels render the unexpanded raceme amentaceous. In the mountain region of the Province of Canterbury, New Zealand, at an elevation from 2500–4000 feet, on open ground and shingles, a variety was collected by Dr. Haast, singular for the narrow-linear leaves of its branchlets.

Dr. Dieffenbach assigns to this plant the native name "Tupakihi," and mentions that the natives use it for dyeing the strings of their mats with a durable black.

CARYOPHYLLEÆ.

COLOBANTHUS BILLARDIERII.

Fenzl in den Annalen des Wiener Museums, i. 48; J. Hook. Flor. Nov. Zeel. i. 27; F. M. Plants of Victor. i. 212; Benth. Flor. Austral. i. 161.

Varietas brachypoda ; leaves channelled, rigid, mostly about ½" long; pedicels hardly as long as or distinctly shorter than the flower.

Chatham-Island.

The same variety has been collected in the Province of Canterbury, New Zealand, by Dr. Haast.

TETRAGONIACEÆ.

TETRAGONIA IMPLEXICOMA.

J. Hook. Flor. Tasman. i. 148, ii. 362; F. M. Plants of Vict. tab. xiii.;
Tetragonella implexicoma, Miq. in Lehm. Plant. Preiss. i. 246.

Widely spreading or frequently *climbing, frutescent;* leaves ovate-rhomboid, occasionally narrow-lanceolate or broad-rhomboid; pedicels axillary, solitary or geminate, downy, longer than the flower, often much longer; lobes of the calyx semilanceolate, nearly twice as long as broad, almost of equal size, inside intensely yellow; stamens generally 12–16; *styles two* rarely three, rather long; ovary two- rarely three-celled; *fruit almost spherical, red, two-seeded* seldom three-seeded, without wings or teeth; putamen small, constricted above the middle, very slightly bilobed at the summit.

Varietas Chathamica; leaves broader, lobes of the calyx more unequal and proportionately broader, styles 3–4 and rather shorter, fruit three- or four-celled.

In damp crevices of rocks on the sea-coast of Chatham-Island.

The normal form of this plant occurs extensively from the west coast of Australia along the whole south coast and on the Tasmanian shores, trailing and climbing, often exposed to the spray of the sea, over sand-ridges and particularly rocks or trees, thus forming not unfrequently graceful festoons. It furnishes as good a spinage as T. expansa.

The calyx is oftener four- than five-lobed. The fruit ripens very scantily. Both T. expansa and T. implexicoma produce always bisexual flowers.

This appears an appropriate occasion to contrast the characters of Tetragonia expansa with those of T. implexicoma, inasmuch as probably both species occur promiscuously in New Zealand and its dependencies, and as the Chatham-plant slightly infringes on some of the characters, by which in Australia both its Tetragoniæ can be easily recognized.

TETRAGONIA EXPANSA.

Murray in Comment. Gœttingens. vi. 13, t. 5 (1783); Scopoli, Deliciæ Faunæ et Floræ Insubricæ, i. 32, t. 14; Solander in Aiton Hort. Kew. ed. i. vol. ii. 178; Thunb. in Transact. Linn. Soc. ii. 335; Willd. Spec. Plant. ii. 1024; Bot. Magaz. 2362; Cand. Plant. Grass. t. 114; Cand. Prodr. iii. 452; Endlich. Prodrom. Flor. Insul. Norfolk. 72; Rich. Voy. de l'Astrolab. i. 320; Payer in Annal. des Scienc. Naturell. 3 serie, xviii. 240–245, pl. 13; J. Hook. Flor. Nov. Zeel. i. 77; Flor. Tasman. i. 147; T. Japonica, Thunb. Flor. Japon. p. 208 (1784); T. halimifolia, G. Forst. Florul. Insul.

Austr. 223 (1786); Plant. Escul. 67; Roth. Bot. Abhandlung. und Beobacht. 48, t. 8; T. cornuta, Gærtner, de Fruct. & Seminib. ii. 483, t. 179 (1791); T. inermis, F. M. in Linnæa, 1852, 384; Demidovia tetragonoides, Pallas, Enumer. Plant. Hort. Procop. a Demidof, t. i. (1781).

Prostrate, *annual* or biennial; leaves deltoid-rhomboid; *flowers subsessile* or on pedicels shorter than the calyx, axillary, solitary or less frequently geminate; *upper lobe of the calyx much broader than long*, semiorbicular or dimidiate-elliptical; the other lobes mostly deltoid, all connivent, slightly yellow inside; stamens usually about 16; styles short, generally 8, occasionally 4–7 or 9; fruit green, almost turbinate, slightly compressed, wingless, 4–9-celled; *seeds generally more than four; putamen usually towards the summit producing four acute teeth*, sometimes unequally eight-toothed or without teeth, at the base excavated, at the summit with as many ridges as fruit-cells.

In Continental Australia sparingly dispersed along the coast from Port Phillip to Moreton Bay; also noticed in the vicinity of Lake Torrens, and in Tasmania; seemingly frequent in New Zealand.

The Australian plant shows no differences from that gathered in New Caledonia by Pancher, at Macao by Hance, in Japan by Th. Siemssen, and in Valdivia by Philippi as compared on this occasion.

MESEMBRYANTHEÆ.
MESEMBRYANTHEMI *sp.*

Found in the Chatham-Islands by Mr. Travers, but the specimina accidentally lost. Most probably the Chatham-plant will prove identical with M. australe (Soland. in Forst. Prodr. 523), since this is the only species hitherto observed in New Zealand and in Norfolk-Island.

LEGUMINOSÆ.
EDWARDSIA GRANDIFLORA.

Salisbury in Transact. Linnean Society, ix. 299; Loddig. Cabinet, t. 1162; Loisl. Herbier de l'Amat. iii. 182; Benth. in J. Hook. Fl. Nov. Zeel. i. 52; E. microphylla, Salisb. l. c.; Bot. Mag. t. 1442; Reichenb. Magaz. der Æst. Bot. t. 19; E. myriophylla, Wenderoth in Linnæa, v. 201; Sophora tetraptera, J. Mill. Icon. Animal. et Plant. t. i.; Meerburgh, Plant. Select. t. 19; Bot. Mag. 167; Lam. Encyclop. t. 325; Red. in Duhamel, edit. nouv. iii. t. 20; Kern. Hort. t. 58; Rœm. Magaz. iv. 10; Sophora microphylla, Ait. Hort. Kew. first edit. ii. 42; Jacquin, Hort. Schœnbr. t. 269; Lam. Encycl. t. 325.

This plant is inserted into this list on the authority of Mr. Travers, who noticed in the bush on the margin of the great lagoon three trees of this plant, about 15' high, but at the time destitute of flowers and fruits.

ROSACEÆ.

POTENTILLA ANSERINA.

Linné, Spec. Plantar. 710 ; Smith's English Flora, ii. 417 ; Engl. Bot. t. 861 ; Lehmann, Historia Potentillar. 71 ; J. Hook. Flor. Novæ Zeel. i. 53 ; Lehm. in Walp. Annal. Botan. Syst. ii. 513, et in Nov. Act. Acad. Cæsar. Leopold. Carolin. xxiii., Suppl. p. 188–191 ; P. anserinoides, Raoul in Annales des Sciences Naturell. trois. serie, ii. 123 ; Raoul, Choix de Plant. de la Nouvell. Zélande, 28 ; Lehm. in Acad. Leop. Carol. p. 191.

On moist places nearly everywhere.

The plant is most probably indigenous neither in the Chatham-Islands nor in New Zealand nor in Australia. In the latter country it is as yet but seen on places widely apart, but gaining ground in its localities.

ANACARDIACEÆ.

CORYNOCARPUS LÆVIGATA.

J. R. & G. Forst. Charact. Gener. Plantar. p. 31, t. 16 ; Linné fil. Suppl. Plant. 156 ; G. Forst. Florul. Insul. Austral. Prodr. 114 ; Willd. Spec. Plant. ii. 1178 ; Lamark, Encycl. Method. t. 143 ; Achill. Richard, Voy. de l'Astrolabe, i. 365 ; All. Cunn. in Annal. of Nat. Hist. iv. 260 ; Bot. Mag. t. 4397 ; Endl. Gen. Plant. 1410 ; J. Hook. Flor. Nov. Zeal. i. 48 ; Benth. & J. Hook. Gen. Plant. i. 425.

Chatham-Island, in woods.

Aboriginal name " Karaka."

The natives use the fruit in the same manner as it is used by the New Zealanders.

The identification of the Chatham-Island plant with that of New Zealand rests solely on a comparison of the leaves.

Dr. Dieffenbach (Journ. of the Roy. Geograph. Soc. of Lond. xi. 206) observes, that the Karaka or Corynocarpus lævigata forms the largest part of the forest of the Chatham-Island, and that it rises to a height of 60', its stem attaining a diameter of 1–3' ; further, that its wood is light and spongy, and that this is the only tree of sufficient size for being used by the natives for making their canoes. A large pigeon finds plentiful food on the fruit of the Karaka-tree.

This noble tree, though rather of slow growth, deserves to be adopted for garden-plantations in all latitudes to which it may adapt itself. The dense dark evergreen of its foliage imparts to the tree a real grandeur. It has well withstood the dry heat of the Australian summer.

ONAGREÆ.

EPILOBIUM TETRAGONUM.

Linné, Spec. Plant. 494; J. Hook. Flor. Antarctica, ii. 270-271; Flor. Nov. Zeel. i. p. 60; Flor. Tasm. i. 117; E. palustre, L. Sp. 495; E. glabellum, Forst. Prodr. 160; E. junceum, Forst. in Spreng. Syst. Veg. ii. 233; E. Billardierianum, Seringe in Cand. Prodr. iii. 41; J. Hook. Fl. Tasm. i. 117, t. xxi.; E. pubens, Ach. Rich. Voy. de l'Astrolabe, 329, t. 36; E. cinereum, A. Rich. l. c.; E. pallidiflorum, Soland. accord. to A. Cunn. in Annal. of Nat. Histor. iii. 31; E. virgatum, E. hirtigerum, E. confertum and E. incanum, A. Cunn. l. c.; E. macranthum, J. Hook. Icon. Plant. 297; E. melanocoulon, Hook. Icon. 813; E. canescens, Endl. Enum. Plant. Hueg. 44; Schlechtend. Linnæa, xx. 646; E. Baueri, Endl. l. c.

In swampy places of Chatham-Island nearly everywhere.

The specimina brought by Mr. Travers approach to the more typical form. The innumerable varieties of this species, which is one of the most protean of the globe, may be collected into three principal groups. One of these groups of varieties comprises Epilobium alpinum, L. Spec. 495; E. origanifolium, Lam. Encycl. ii. 376; E. haloragifolium, A. Cunn. in Annal. Nat. Hist. iii. 34; E. microphyllum, A. Rich. Voy. de l'Astrolabe, i. 325, t. 36; E. confertifolium, J. Hook. Flor. Antarct..i. 10, et Icon. Plant. 685; E. tenuipes, J. Hook. Fl. Nov. Zeel. i. 59.

Another complex of varieties, more particularly though probably not absolutely restricted to New Zealand, is formed by E. rotundifolium, Forst. Prodr. 161; E. nummularifolium, E. pedunculare, E. nerteroides, E. alsinoides, E. thymifolium, E. atriplicifolium, All. Cunn. in Annal. of Nat. Hist. iii. 31 et seq.; E. linnæoides, J. Hook. Flor. Antarctic. i. 9, t. 6; E. macropus, Hook. Icon. Plant. t. 812.

All these plants with a host of other synonyms, which it would be superfluous to conjoin on this occasion, might when contrasted with the very few other well-marked species of this genus, such as E. latifolium, E. luteum, E. suffruticosum and E. hirsutum, be recognized by the following brief diagnosis : Flowers axillary and terminal, solitary, not truly racemose ; lobes of the calyx connate towards the base ; petals bilobed, pink or pale ; stamens and style included and almost straight ; filaments not distinctly dilated at the base ; anthers small ; style glabrous ; *stigma club-shaped ;* ovules uniseriate ; capsules narrow-cylindrical ; coma several times longer than the seeds.

Dr. Hooker first recognized the wide distribution of E. tetragonum over the globe (conf. Flor. Antarctic. ii. 270). The author of this memoir, in watching the Epilobia for many years in Australia, is persuaded, that no other but this species occurs in this continent, where it is widely dispersed over the extratropical part, reaching tropical latitudes towards the east coast, ascending to the summits of the alps and penetrating to the edges of the arid deserts. Adapting itself to so varied climatic conditions the plant assumes corresponding protean forms, though, as remarked, apparently not those forms which in the humid climate of New Zealand have their prototype in E. rotundifolium. Some of the extreme varieties of the species appear to the isolated observer to indicate specific distinction; but it would be a vain attempt to draw the limits between these forms, as no diagnosis founded on them would stand the field-test. Through the kindness of W. Woolls, Esq., of Parramatta, I received specimina, collected by All. Cunningham at the Bay of Islands in July 1838, and to which Cunningham attached the name E. rotundifolium, thus apparently discarding the distinctions on which his E. nummularifolium, &c., rested; for these specimina accord more closely with Dr. Hooker's definition of E. nummularifolium than with that of E. rotundifolium. The copious production of seeds of all Epilobia renders it unlikely, that specific demarcations between them have been obliterated by hybridism. It is a curious fact that seemingly hitherto no other Epilobia have been found in the southern hemisphere than those referable to E. tetragonum.

CORNEÆ.

COROKIA BUDDLEYOIDES.

All. Cunn. in Annal. of Nat. Hist. iii. 249; Hook. Icon. Plant. 424; J. Hook. Flor. Nov. Zeel. i. 98.

Common in the woods of Chatham-Island.

A stately plant, dwarf when growing on the sea-coast. Leaves lanceolate-ovate or lanceolate, without conspicuous stipules. Bracteoles at the base of the pedicels canalicular-lanceolate, $\frac{1}{2}$-1′′′ long. Teeth of the calyx 5, occasionally 6 or 7, deltoid, about $\frac{2}{3}$′′′ long. Petals semilanceolate, with broad base sessile, 2-2$\frac{1}{4}$′′′ long, inside yellow, tardily dropping. Stamens inserted outside of the narrow epigynous disk. Anthers yellow, ellipsoid, dorsifixed, hardly 1′′′ long. Ovary two-celled, each cell containing a single pendent ovule. Fruit a true drupe, with orange-colored pericarp. Putamen egg-shaped, thick, bony, one-celled. Seed about 2′′′ long, ellipsoid-cylindrical. Testa tender-membranous, pallid. Embryo equally slender-cylindrical, nearly as long as the albumen, straight; the cotyledons semielliptical, several times shorter but not thicker than the superior radicle.

This plant is strictly referable to Corneæ, a group seemingly only discernible from Araliaceæ by its bony putamen and perhaps the length of the embryo. Koch (Synops. Flor. German. i. 353) records wrongly the embryo of Hedera Helix about as long as the albumen. Bennett in his learned remarks on Polyosma (conf. Horsf. Plant. Javan. rarior. 194) reduces the Alangieæ to the family of Corneæ, which thus amongst Australian and New Zealandian plants would comprise, Corokia buddleyoides, Corokia Cotoneaster, Griselinia lucida, Rhytidandra polyosmoides (F. M. Fragm. Phyt. Austr. ii. 84 & 176).

Polyosma Cunninghami, according to R. Br. (in Horsf. Plant. Javan. 196) is rather referable to Escallonieæ than to Corneæ. Recent discoveries of new types of genera have, however, much invalidated the formerly assumed characteristics of all allied orders. Thus the genus Mackinlaya (F. M. Fragm. Phyt. Austr. iv. 119) renders the transit from Araliaceæ to Umbelliferæ almost complete.

EUPHORBIACEÆ.

EUPHORBIÆ sp.

A plant of this genus was observed in the Chatham-Islands by Mr. Travers. It is most likely Euphorbia glauca (Forst. Prodr. 208 ; Achill. Rich. Voy. de l'Astrolab. 352 ; J. Hook. Fl. Nov. Zeel. i. 227), a plant common in New Zealand and to be found also in Norfolk-Island.

UMBELLIFERÆ.

GINGIDIUM DIEFFENBACHII.

Tall, erect ; sheaths of the petioles produced at the summit in two blunt teeth ; *leaves thrice pinnatisected ; their segments long, linear, flat, flaccid,* hardly or slightly spreading, seven-nerved, short-mucronate ; rachis and rachillæ imperfectly jointed ; floral leaves with generally five segments and a petiole vaginate to the summit ; panicle consisting of numerous pedunculate compound umbels ; *leaflets of the general and special involucres very few,* narrow-lanceolate ; teeth of the male calyx short.

On damp places under cliffs at the sea-shore.

Of this seemingly hitherto undescribed plant the collection contains leaves and a portion of the aged male inflorescence. From

c

this material it appears that the species approaches closely to Gingidium antipodum (Anisotome antipoda, J. Hook. Flor. Antarctic. i. 17, t. ix.); it differs, however, in longer not so much spreading evidently flat segments of the leaves, in longer teeth terminating the petiolar sheaths, in not numerous leaflets of the involucres and involucels, in less lengthened teeth of the male calyx and possibly in its carpological characters, which remain as yet unascertained.

Nevertheless this and all the allied plants, including Gingidium squarrosum, G. Monroi, G. glaciale, G. procumbens, need yet a more extensive scrutiny in the field, before their specific value can be ascertained and their respective diagnoses can be absolutely fixed.

The lower petioles of G. Dieffenbachii are free of a vaginal membrane towards their summit. The segments of the leaves are about 1''' broad or slightly broader, and the majority attains a length of from 2-3". The leaflets of the general involucre are from 4-8''' long, are like those of the involucels placed somewhat unilaterally and number often only three. The teeth of the male calyx are deltoid and semilanceolate and generally less than ½ line long.

The name of Dr. Ernest Dieffenbach is attached to this plant for commemorating that we owe to this philosopher the first record of the indigenous natural productions of the Chatham-group.

Gingidii sp.

Mr. Travers collected on grassy places of Chatham-Island a second Gingidium, to which, should it prove new, the name G. Traversii might be given. It seems sufficiently characterized by leaves with a but minutely denticulated sheath and (as far as the material of the collection admits to judge) with two or three rigid segments, which are somewhat broader than those of the foregoing species and streaked by many longitudinal nerves; the floral leaves are simply linear and undivided, passing into the petiole without any denticulation. The leaflets of the general and special involucres are more numerous and narrower than those of G. Dieffenbachii; the petals roundish, almost sessile and fully as broad as long. It seems nearest in its affinity to G. Monroi.

RUBIACEÆ.

Coprosmæ sp.

In Mr. Travers's collection are contained two species most probably of this genus from the Chatham-Island; these can seemingly not be referred to any congeners of New Zealand ; but as both kinds at the time of Mr. Travers's stay on the islands were devoid of flowers and fruit, the characteristic of these plants can only be given on a future occasion.

COMPOSITÆ.

EURYBIA TRAVERSII.

(Sect. Cardiostigma.)

Arboreous ; leaves short-petioled, large, flat, opposite, broad- or lanceolate-ovate, perfectly teethless, beneath as well as the branchlets and peduncles pale-silky, above smooth and shining ; the primary veins forming large meshes ; the veinlets but slightly conspicuous ; *panicles cymose, axillary and terminal, subsessile, with short opposite branches and branchlets,* condensed into a leafy rich inflorescence ; *capitula small, without rays,* 5–16-flowered ; scales of the involucre 6–13, acute, outside as well as the peduncles and bracts silky ; corollæ appressed-downy ; those of the female flowers half as long as the male ones, considerably shorter than their style and their pappus, attenuated into a minute ligular apex ; corollæ of the bisexual flowers little longer than the pappus ; stamens enclosed ; *stigma of the bisexual flowers with extremely short lobes ;* pappus twice or less than twice as long as the almost silky achenia ; its outer bristles half or nearly as long as the inner ones, all slightly scabrous.

Generally distributed through the woods of Chatham- and Pitt-Island, still most abundant near the sea-border.

A very beautiful not viscid tree, attaining a height of 30–35', called inappropriately by the colonists Bastard Sandalwood-tree and passing under the native name "Ake-Ake." The stem often 4' in girth, but almost always hollow, a character it has in common with Eurybia argophylla. Branches and branchlets opposite ; the latter somewhat quadrangular, and covered with a very close indument. Petioles 2–5''' long. Leaves coriaceous, 1½–2½" long, 1–1½" broad, rather distantly and spreadingly nerved, tapering at the base, minutely apiculate at the generally somewhat acute apex. Panicles towards the summit of the branches rather copiously axillary, by the fall of the leaves construing a terminal inflorescence. Ultimate peduncles often shorter than the capitula and occasionally suppressed. Bracts at the base of the general peduncle imbricate ; those at the base of the succeeding and other peduncles solitary, opposite, 1–3''' long, narrow-lanceolate, all persistent. Capitula from semiovate to hemispherical. Scales of the involucre in very few rows, 1–1½''' long, narrow- and linear-lanceolate. Female flowers circumferential; their corolla only about 1''' long, forming a very slender cylinder, which terminates in an entire or slightly slit apex of much less length than the tube; style less than half exserted; stigmata about ⅓''' long. Bisexual flowers with a corolla of nearly 2''' length, which is campanulate and five-toothed at the summit; style short-exserted; stigma terminated into two roundish lobes of only about ⅛''' length. Anthers nearly ¾''' long. Achenia ½–1'" long, prominently lined. Longest bristles of the pappus 1–1½''' long; all fulvid.

On this, the noblest plant brought by Mr. Travers from the Chatham-group, I gratefully bestow the name of the young promising naturalist and of his enlightened father, though offering thereby but a faint acknowledgment of their generous response to a proposition of adding by a phytological exploration of the Chatham-Islands a limited but interesting contribution to the physical geography of the globe.

Eurybia Traversii counts amongst the very few species of the vast order of Compositæ, attaining the height of stately trees. In Continental Australia and Tasmania we know only Eurybia argophylla to attain the same and occasionally even greater dimensions; indeed the latter may be the most gigantic species of this order anywhere in existence, like amongst Labiatæ our Prostanthera lasianthos, which perhaps represents in that order the greatest if not only timber tree. Eurybia argophylla attains occasionally the height of 60', and in this regard amongst coordinal plants only Synchodendron ramiflorum of Madagascar and Melanodendron integrifolium of St. Helena seem to rival with it. Eurybia lirata and Bedfordia salicina may be noted 30' high, but their stems are never so robust as those of the forenamed plants; and none of the tall Eurybia and Senecio species of New Zealand seem even to equal these latter in height, unless Senecio Forsteri.

Eurybia Traversii is to be drawn to the series of species with always or frequently opposite leaves, apparently not represented in New Zealand, but of which from Continental Australia the following are known: E. viscosa (Cassini Dict. xxxvii. 487), E. chrysophylla (Cand. Prodr. v. 266), E. megalophylla (F. M. in Papers of the Roy. Soc. of Tasm. iii. 228), E. alpicola (F. M. l. c. 229), E. rosmarinifolia (Cand. Prodr. v. 268), E. oppositifolia (F. M. Fragm. Phyt. Austr. ii. 88). From all these the Chatham-plant is well distinct by its tall arboreous growth, by the form of its leaves, by its inflorescence, and particularly by the singular reduction of the size of the female flowers, those however of E. oppositifolia being as yet unknown. In the latter character E. Traversii responds to some extent to the species collected into the section Brachyglossa by Candolle and still more intimately to E. tubuliflora (Sond. & Muell. in Linnæa, xxv. 455), in which the minute female corolla is also decidedly shorter than the style. The latter plant, from the vicinity of St. Vincent's Gulf, is however in other respects widely different.

With E. argophylla the Chatham-plant has beyond its habitual grandeur many characters in common; but except other notes the alternate toothed leaves clothed beneath with a different indument, the more elongated differently branched panicles, the conspicuous development of ligules of the female flowers distinguish at once E. argophylla. Perhaps a nearer approach is afforded by E. Beckleri, an as yet undescribed species, discovered by Dr. Beckler on the Clarence-River, similar in position, size and form. of leaves to E. argophylla, distinguished however by their more velvety than silky indument and their toothless margin; of this the flowers and fruits are as yet unknown, but the remnants of involucres indicate seemingly sufficiently its generic position.

Possibly from all congeners E. Traversii recedes by its but very slightly divided stigma of the female flowers. Still in other genera of Compositæ this character has proved so little fixed, that only sectional value is attached to it on this occasion.

Living plants, caused some years ago to be sent from the Chatham-Islands by the Honorable Dr. Featherstone, Government Superintendent of the Province of Wellington, N. Z., to the Botanic Garden of Melbourne, have borne as yet no flowers.

Mr. Will. Seed, Landing Surveyor of Wellington, refers in a printed report on the capabilities and productions of the Chatham-Islands, dated 21st October 1861, also to the Ake-Ake as amongst the principal woods of this island-group.

EURYBIA SEMIDENTATA.

Olearia semidentata, Decaisne, Planch. Voy. Venus accord. to J. Hook. Flor. Nov. Zeel. i. 114.

Leaves alternate, lançeolate, flat, above the middle distantly serrulate, beneath thinly white velvet-downy, tapering into a sessile base ; peduncles araneous-woolly, crowded towards the summit of the branchlets, often considerably longer than the flower-heads, beset with several small entire lanceolate leaves, terminated by a single capitulum of copious flowers ; involucre consisting of numerous linear-lanceolate towards the summit araneous scales, of which the outer ones are half or more than half as long as the inner ones; corollæ blue; ligules of the female flowers 17–22, entire, at least twice as long as the tube; their style enclosed; bisexual flowers about as long as the involucre; their corolla almost glabrous, of the length of the pappus; their anthers enclosed; their stigmata about one-third the length of the style; pappus little or hardly longer than the slightly pubescent achenium; its outer bristles variously shorter than the inner ones.

In damp open places of Chatham-Island.

Probably only of shrubby growth. Leaves on the summit of the branchlets, 1–2″ long, above soon glabrescent and shining, 3–5-nerved, thinly net-veined; the midnerve more prominent, the lateral ones faint. Peduncles 3″ or less long; their leaves measuring 3–6‴. Scales of involucre 2–3½‴ long. Ligules 3–4‴ long, oblong-lanceolate. Corollæ both of disk and ray (according to Mr. Travers) vividly blue; those of the disk above the middle bell-shaped, at the summit five-toothed. Anthers short-caudate. Style of either kind of flower enclosed. Stigmata exserted, hardly ¼‴ long, narrow semiterete-linear. Achenia 1¼‴ long, streaked. Pappus pale, not thickened towards the apex.

Mr. Travers's collection contains an allied plant, in all probability not claiming specific distinction. It differs in blunt-toothed larger leaves, attenuated more distinctly into a broad petiole, covered beneath as well as the peduncles with a very dense tomentum, further in larger capitula of which the partial vestiment is much thicker though also araneously coherent, in still more numerous and larger flowers and in a more evidently downy achenium.

At the edges of woods near the sea-shore not common.

The capitula enclose often about thirty ligules, which according to Mr. Travers assume from a deep blue gradually a whitish hue, and from 70–80 discal flowers, also blue but persistently so.

This plant is evidently allied to Eurybia operina (Arnica operina, Forst. Prodr. 299 ; Olearia operina, J. Hook. Fl. Nov. Zeel. i. 114), which differs in larger more deeply crenate leaves, longer peduncles, larger capitula, yellow discal flowers (according to Forster & Ach. Rich. l. c. 233). The absence of authentic specimina of the New Zealand plant in the Melbourne collection prevents the writer to carry the comparison further on this occasion. The leaves of the larger variety of the Chatham-plant are not unlike those of Eurybia Colensoi (Olearia Colensoi, J. Hook. Flor. Nov. Zeel. i. 115).

Differences, such as are pointed out as existing between these two Chathamian plants, may be noted in several Australian species.

SENECIO HUNTII.

Arboreous, clammy; branchlets and peduncles glandulous-downy; *leaves lanceolate, entire, firm, reflexed at the margin, from subtle downs pale beneath, with unenlarged base sessile; above gradually glabrescent;* their veins immersed; *panicle compact, terminal,* surrounded by leaves, somewhat pyramidal, producing numerous flower-heads; peduncles copiously beset with very short brown gland-bearing hair; *capitula ligulate, with very numerous flowers;* involucre from semiovate verging to semiglobose, supported by few linear-subulate bracts; its scales about 13, unequal, mostly blunt, not much shorter than the discal flowers, somewhat glandular-downy; alveoles of the receptacle toothless; ligules of the female flowers not much longer than their tube, entire; anthers almost entirely exserted; bristles of the pappus nearly as long as the discal corollæ, twice or thrice as long as the glabrous achenia, not thickened towards the apex, almost biseriate.

On damp localities of woods growing generally in patches; rare in Chatham-Island, common in Pitt-Island.

A tree, often attaining according to Mr. Travers a height of 25′, called "Rautine" by the aborigines. Branchlets cicatricose, seemingly soon defoliated. Leaves, as far as seen, about 3″ long, ⅔-1″ broad, with spreading primary and closely netted immersed secondary veins. Panicle devoid of any long universal peduncle, interspersed with some generally short and narrow floral leaves. Ultimate peduncles somewhat or hardly longer than the capitula. Bracts near the apex of the special peduncles and around the involucres 1-2‴ long. Scales of involucre 2½-3‴ long, linear- and lanceolate-oblong, bearded at summit, imperfectly downy at the back, finally separating from each other. Ray-flowers 15-18; their ligule 2-3‴ long, lanceolate-oblong, yellow; their style partially exserted. Disk-flowers about 40; their corolla barely 3‴ long, campanulate above the middle, five-toothed at the summit. Anthers about 1‴ long. Stigmata exserted. Achenia measuring about 1‴ in length, furrowed, slender. Pappus white, very tender, composed of 50-60 serrulate indistinctly biseriate bristles.

This plant received its specific signification in order that the name of Mr. Frederick Hunt may also phytologically for ever be identified with that of the small isle, of which he was the first and is still the principal European occupant, and in which this remarkable species forms such a prominent feature of the primeval vegetation. Mr. Hunt is moreover highly entitled to this mark of respect for the kind assistance which he afforded to the young traveller in his exertions of rendering known the vegetable products of these islands. A hope is simultaneously expressed, that as a permanent resident there the hospitable settler of Pitt-Island may also hereafter advance our cognizance of the vegetation around him.

In its arborescent growth Senecio Huntii has probably amongst the many hundreds of its congeners in the Victorian and Tasmanian Senecio Bedfordii (Bedfordia salicina, Cand. Prodr. vi. 441) and in the New Zealand S. Forsteri (J. Hook. Fl. Nov. Zeel. i. 148, t. xl.) its only rivals. In deep humid forest-gullies, favorable for its luxuriant development, S. Bedfordii equals S. Huntii in height, and thus both excel as the tallest all others of a cosmopolitan genus, which is recognized next to Solanum as the richest of all in species.

In the systematic series S. Huntii may perhaps find its place nearest to the exclusively Tasmanian S. Brownii (Centropappus Brunonis, J. Hook. in Lond. Journ. of Bot. vi. 124 ; Flor. Tasman. i. 225, tab. lxv.), which although much smaller exceeds also most other species in size. Both plants have many characters in common, but the Tasmanian plant differs in its smoothness, narrower flat leaves, smaller panicles with less large capitula and often shorter peduncles, much less numerous flowers of the heads, fewer and shorter scales and broader and shorter bracts of the involucre as also more grossly serrated upwards thickened bristles of the pappus.

S. Huntii approaches likewise in many characters to S. glastifolius (J. Hook. Fl. Nov. Zeel. i. 147, tab. xxxix.), which recedes again in its smoothness, in petioled often toothed leaves, in a less compact inflorescence with fewer and larger capitula, in larger bracts and also in upwards somewhat thickened pappus-bristles.

The transit of Bedfordia and Centropappus to Senecio, indicated by Dr. J. Hooker, is rendered sufficiently clear by the discovery of this Chathamian species.

SENECIO RADIOLATUS.

Upper leaves above the middle toothed and often also pinnatifid, below the middle entire, with cordate or auriculate base sessile, beneath somewhat arachnoid-downy ; their lobes almost semilanceolate; flower-heads numerous, corymbose-paniculate; *receptacle pitted with toothed alveoles;* involucres from cylindrical gradually somewhat bell-shaped, consisting of 12–16 unequal scales, surrounded by several short subulate-semilanceolar bracts; flowers 60–70, not exserted; corolla of bisexual flowers but little wider towards the summit; anthers enclosed; *ligules of the female flowers minute, slightly shorter than their tube,* considerably stretching beyond the stigmata; achenia slender, not half as long as the involucre, grey from a thin appressed when moistened mucous indument, of about half the length of the very tender pappus; the latter hardly emerging beyond the involucre.

Chatham-Island on sandy places at the sea-beach.

The collection contains flowering branches; it remains therefore to be ascertained whether the plant is annual, herbaceous-perennial or frutescent. Branches rather stout, not woody, glabrescent. Upper leaves flat, herbaceous, in circumference approaching to ovate- or lanceolate-oblong, from the middle downward generally slightly dilated; the upper portion of the leaves always acutely toothed and not unfrequently deeply cut into lobes. Inflorescence terminal, consisting of very many capitula. Peduncles 1″ or less long; most of the special peduncles longer than the flower-heads; all bearing several subulate- or lanceolate-linear bracts. Involucre 3–4‴ long; its scales linear and lanceolate-linear, glabrous except at the tip, forming almost two rows, the narrower outer ones generally overlapping the scarious margins of the inner scales. Basal bracts of the involucre but simply sphacelate at the apex. Bisexual corollæ about 2‴ long. Radial flowers 10–14; their ligules hardly longer than 1‴, oblong-lanceolate, generally toothless, soon rolled back. Achenia scarcely above 1‴ long, only about ½‴ thick; their indument expanding into a transparent mucus when exposed to warm water.

This species seems to find amongst New Zealand plants its nearest allied in S. latifolius and S. Banksii. The former of these recedes, according to Dr. Hooker's notes, in larger leaves, a glandular indument of the achenia and long-emergent pappus. S. Banksii differs (Fl. Nov. Zeal. i. 145) in smaller capitula and achenia almost as long as the involucre.

Should S. radiolatus prove annual it might be drawn into the series of S. vulgaris. The latter, which has immigrated into several parts of Australia, is principally distinguished by its less robust branches, smaller leaves, which are toothed in their whole length and cut into shorter and blunter lobes, less numerous capitula with fewer flowers, narrower scales of the involucre, the comparative smoothness of the receptacle, which is not raised into toothed

D

alveoles, corollæ at least in ordinary forms if not always uniform-tubular and bisexual, achenia lined with rows of subtle indument, which deliquesces to a less copious mucus when moistened. Nearly the same characters distinguish S. silvaticus and S. lividus from the Chatham-plant; but these two plants stand to it in a still nearer relationship on account of the extremely short ligules of their female flowers, which rays are however generally not half as long as their tube. S. viscosus recedes moreover in glabrous achenia; its ligules are half or nearly as long as the tube; otherwise it differs but faintly from S. silvaticus and S. lividus, from which also S. crassifolius and S. vernalis again mainly seem to differ only in broader and still longer ligules. How far the short-ligulate of these plants are claiming distinct positions in the botanical system a renewed critical disquisition must show; and it will be well for the phyto-grapher, who may attempt such disquisition, to be conscious how multifarious an array of forms S. lautus presents in Australia.

Amongst native Australian plants only one is known with yellow ligules so short or even shorter than those of S. radiolatus, those of S. leucoglossus (F. M. Fragm. Phyt. Austr. ii. 15) being whitish. This so far allied plant is one, distributed from the author's Austra-lian collections about sixteen years ago as S. BRACHYGLOSSUS, and subsequently described by Dr. Sonder as Erechtites glossanthes (Linnæa, xxv. 524). It is not strictly referable to the section Pla-giotome of Erechtites, inasmuch as the ligules are considerably longer than the style, and it cannot be far separated in a systematic arrangement from S. silvaticus and its allies, with which it shares the annual duration and very short ligules, the latter being half or more than half as long as the tube, varying in length from ⅓–1‴. It adds thus another form, by which the generic limitation of Erech-tites is rendered quite untenable. S. brachyglossus, for which, should a plant of the same name since published by Turczaninow (Bullet. de la Soc. Impér. de Mosc. xxiv. 2, 87) prove well defined, the name of S. BREVILINGUEUS may be substituted, is readily recognized from S. silvaticus in either lobeless or acutely lobed leaves with less or not amplexant base, in shorter tubes of the female flowers, in glabrous often shorter and fewer involucre-scales, in capitula with fewer flowers, in dense- or even velvet-downy achenia and in a shorter pappus. These characters it maintains in widely distant localities. It has been found by the author originally near St. Vincent's Gulf and subsequently at Point Nepean and elsewhere about Port Phillip, also on the Murray-River; by Dallachy on the Wimmera, by Beckler on the Darling-River, by Wheeler near Stokes-Range and Cooper's Creek, by Burkitt on the Lachlan-River and by Oldfield on the Murchison-River.

LEPTINELLA FEATHERSTONII.

Cotula Featherstonii, F. M. coll.

Stems branched, short-downy, *not creeping; leaves obovate- or oblong-cuneate, with exception of the three-toothed summit entire, tapering into a sessile base,* invested with subtle downs ; peduncles axillary and terminal, less than thrice as long as the capitula, densely short-downy ; capitula with numerous female and very many bisexual flowers ; scales of the involucre 10–12, nearly biseriate, unequal, almost semilanceolate, thinly downy outside ; receptacle depressed, hemispherical ; female flowers pedicellate ; their corolla minute, pale, glandless, nearly conical, about half as long as their achenium ; corolla of bisexual flowers four-toothed, glandless ; their anthers exserted ; fertile achenia obovate-cylindrical, streaked, slightly glandulous.

On damp rocks near the sea in one locality on the south-east side of Chatham-Island.

A singular plant, according to Mr. Travers of luxuriant growth, entirely dissimilar in habit to all other Leptinellæ, and in general appearance more resembling the species of Myriogyne. Two pieces of a ramified stem or its branches, sent by Mr. Travers, seem sufficiently to indicate a robust erect or at least not radicant growth of this species. Branches stout, grey from short soft somewhat appressed hair. Leaves alternate, flat, $\frac{3}{4}$–$1\frac{3}{4}$″ long, towards the summit $\frac{1}{4}$–$\frac{1}{2}$″ broad, at the blunt apex by generally two extremely short incisions three-toothed, occasionally almost toothless, tapering into a somewhat petiolar base, of equal color on both sides and also clothed on both pages with subtle downs, finely three-nerved, reticulated by immersed hardly perceptible veins, ciliate by dense but extremely short downs. Peduncles monocephalous, 1″ or less long. Scales of involucre $\frac{3}{4}$–2‴ long, outside short-downy. Capitula depressed-globular, about 4‴ broad, of strong but not unpleasant odor. Female flowers numerous, peripherical in many rows ; corollæ $\frac{1}{3}$–$\frac{1}{2}$‴ long, broadly oblique-conical, glabrous, indistinctly toothed at the contracted apex ; style enclosed or its apex exserted, lobes of the stigma exceedingly short ; achenia streaked by several prominent lines, not compressed, hardly longer than half a line, brown-grey ; their pedicel half or less than half as long as the achenium, the latter readily separating, leaving the receptacle closely covered with these persistent stipites. Male flowers central, numerous, developing no achenium ; corollæ about 1''' long, yellowish. Anthers 4, pale-yellow, $\frac{1}{6}-\frac{1}{4}'''$ long, acute at the apex, without appendages at the base. Style simple, enclosed. Stigma truncate, very slightly divided in the middle.

It is not without hesitation that the author draws this remarkable plant to Leptinella. Had the analysis revealed any appreciable structural differences in the organization of the flowers from those of the complex of the known species, our plant would as *Traversia* have been raised to generic distinction. But to the mere habitual disresemblance of this new member of the genus Leptinella can hardly more value be attached than to the external dissimilarity of Cotula australis to C. coronopifolia. From Myriogyne, a genus to which Sphæromorphæa must be reduced, it differs already in the want of the development of achenia of the bisexual flowers and in their simple stigma. The genus Trineuron, to which Ceratella might be joined, stands still more distant. Whether, however, it would not be preferable to bring all Leptinellæ to Cotula, to which some species, indicated by C. Koch under Pleiogyne (Schlecht. & Mohl, Bot. Zeitung, i. 40) offer a transit, must by a future careful revision of all existing allied plants be decided.

The writer selected this plant, one of the most interesting of those of the Chatham-Islands, for bearing the name of the Honorable Dr. Featherstone, Government-Superintendent at Wellington, New Zealand, as a mark of recognition of the favor shown by that able officer in introducing some years since a number of living plants from the Chatham-Islands into the Melbourne Botanic Garden.

It would be desirable to ascertain the duration of this species.

Leptinella potentillina.

Stems creeping, soft-downy ; petioles towards the base membra-nous-expanded and clasping; leaves pinnatisected, finely gland-dotted, almost glabrous ; *segments of the leaves subovate, with duplicate short teeth* or minute toothed lobules; peduncles axillary, many times longer than the capitula, soft-downy ; capitula with very numerous sessile flowers ; *scales of the involucre about thirty,* arranged in several rows, almost smooth ; the outermost semilan-ceolate, acute ; the others semioval and cuneate-oval ; receptacle hemispherical ; *corollæ of the female flowers almost semiovate, about half as long as the achenium;* corolla of bisexual flowers four-toothed, glandless; their anthers almost enclosed; fertile achenia clavate-cylindrical, slightly glandulous. On swampy places of Chatham-Island. Stems rooting. Leaves in circumference obovate or elliptical, generally 2–3″ long, with a petiole of about 1″ length; segments generally approximate or the lower ones somewhat remote, flat; the upper ones 3–5‴ long, the others gradually smaller; the marginal incisions short and rather unequal; the teeth acute. Peduncles 2–3″ long, solitary. Capitula strongly odorous. Scales of involucre 1–1½‴ long, except the outermost blunt. Receptacle closely dotted by the rudimentary pedicels. Female flowers in many rows, somewhat more numerous than the bisexual ones; the corolla of the former about ⅓‴ long; their style almost entirely enclosed; the minute bilobed stigma exserted; the achenia ½–¾‴ long. Corolla of the bisexual flowers nearly 1‴ long, obconic-cylindrical, glabrous; their style undivided and almost or entirely enclosed; anthers about ¼‴ long, acute at the apex, not caudate at the base; the achenia ¼‴ or less long.

Leptinella scariosa (Cassini, Bullet. Philomat. 1822, 127; L. acænoides, Hook. & Arn. in Hook. Journ. of Bot. iii. 325) is distinguished by the paucity and orbicular form of the involucre-scales, by the almost uniseriate female flowers and the different shape of their corolla.

Leptinella lanata (J. Hook. Flor. Antartic, i. 26, t. 19) differs in its woolly indument, in its much smaller leaves with two- or few-toothed segments, in shorter densely woolly peduncles, gland-dotted corollæ, of which the peripherical ones are longer than their achenium and of an ovate-cylindrical form.

Leptinella plumosa (J. Hook. Fl. Antartic. i. 26, t. 20) recedes in a more copious almost silvery indument, often larger leaves, the segments of the latter deeper dissected and their secondary lobes slit into more narrow and acute teeth, in less numerous though not uniseriate involucre-scales, in female corollæ about twice as long as broad and rather longer than the achenium and in five-toothed bisexual flowers.

Leptinella propinqua (J. Hook. l. c. i. 27) approaching L. potentillina nearest, is recognized by an again denser woolly indument, more divided segments of the leaves with acuter lobes, less numerous involucre-scales and gland-dotted corollæ, and perhaps also by the structure of flowers and fruits.

L. squalida (J. Hook. Fl. Nov. Zeel. i. 129) is known by its silkiness, leaves lengthened into a narrow-oblong outline with recurved narrower segments, scapes longer than the leaves and probably also in some of its floral characters.

L. minor (J. Hook. l. c.) recedes again in leaves with narrower outlines and segments, in glandulous flowers, of which the outer ones are pedicellate and in less numerous scales of the involucre.

L. dioica (J. Hook. l. c.) differs in less deeply divided glandless leaves, distinct male and female capitula.

L. pusilla (J. Hook. l. c.) in smaller and narrower leaves with sometimes entire lobes, an involucre of very few and therefore almost uniseriate scales, and seemingly dioecious capitula with only about 6–8 male flowers.

L. Hookeri (L. multifida, L. intricata, L. longipes, J. Hook. in Lond. Journ. of Bot. vi. 117–118) is recognized by its comparative smoothness, by leaves with fewer scantily toothed segments, an involucre with very few biseriate scales of roundish form, capitula with fewer flowers, distinctly glandular female corollæ, which in form however approach to those of L. potentillina.

L. filicula (J. Hook. Flor. Tasm. i. 194), an alpine species, differs in smaller leaves with fewer scantily toothed segments, short peduncles, glandular corollæ, fewer and oval involucre scales.

One species from the Province of Canterbury, New Zealand, apparently not recorded in Dr. Hooker's flora, and of which our museum owes specimina to Mr. Travers senior and Dr. Haast, is remarkable for the glandular and entirely black flowers, producing a slender cylindrical distinctly toothed female corolla.

TARAXACUM OFFICINALE.

Weber in Wiggers's Primitiæ Floræ Holsaticæ, p. 56 (1780); Koch, Synops. Flor. German. et Helvet. ii. 492; J. Hook. Flor. Nov. Zeel. i. 152; T. dens leonis, Desfontain. Flor. Atlantic. ii. 228 (1800); Candolle, Prodr. vii. 145; J. Hook. Flor. Antarct. ii. 223, tab. cxii.; Leontodon Taraxacum, L. Sp. Plant. 1122; Smith, Flor. Britan. 823; Smith, English Flora, iii. 350; Leontodon palustre, Lyons, Fascic. Plant. Cantabr. 48.

On open land of Chatham-Island.

This plant is also immigrated into Australia and Tasmania, where it gradually spreads onward. Dr. Milligan collected it on Flinders-Island and on other of the isles of Bass-Straits. Dr. Haast has brought this plant even from the glacier-regions of Middle-Island, New Zealand.

Sonchus oleraceus.

Linn. Spec. 1116; Forst. Prodr. 282; R. Brown, Append. to Flinders's Voy.
592; Ach. Rich. Voy. de l'Astrolabe, i. 230; J. Hook. Fl. N. Zeel. i. 153;
S. asper, Fuchs, Histor. 674; J. Hook. Fl. Tasm. i. 227.

On Pitt-Island only, where it grows on the sides of cliffs at the
sea-coast.

Mr. Travers's plant exhibits a very luxurious variety and has its
peduncles densely covered with closely appressed white wool. As
far as the material permits of judging the Chatham-plant is not
distinct from the Linnean species, but the fruit requires yet to be
compared.

LOBELIACEÆ.

Lobelia anceps.

Thunberg, Prodrom. Flor. Capens. p. 40; Rich. Voy. de l'Astrol. 227; Alph.
de Candolle in Cand. Prodr. vii. 375; J. Hook. Flor. Nov. Zeel. i. 158;
L. alata, Labill. Plant. Nov. Holl. Specim. i. 51, t. 72; R. Br. Prodr. 562;
Ræm. & Schult. Syst. Veg. v. 69; Endlich. Prodr. Flor. Insul. Norfolk,
50; L. cunciformis, Labill. l. c. 51, t. 73.

On swampy or damp rocky places of Chatham-Island.

The stems of this plant are occasionally quite prostrate and
radicant. The upper lobes of the corolla are linear, channelled,
ascendent and outside purplish-brown; the lower lobes oblong-
cuneate, horizontal, blunt, apiculate, with an outside prominent
purple costa; the faux at the lower lip is white; the tube inside
yellowish.

SOLANACEÆ.

Solanum aviculare.

Forster, Prodrom. 107; Forst. Plant. Escul. 12; Achill. Richard, Voyage de
l'Astrolabe, 193; J. Hook. Flor. Nov. Zeel. i. 182; F. M. in Transact.
Victor. Inst. 68; S. lacinatum, Aiton, Hort. Kew. first edit. i. 247; Willd.
Spec. Plant. i. 1030; R. Brown, Prodrom. 445; Botan. Magaz. 349;
Botan. Cabin. 717; Ræm. & Schult. Syst. Veget. iv. 578.

Common near edges of creeks of Chatham-Island.

The plant extends in Continental Australia from the eastern
boundary of the colony of South Australia to the southern part of
Queensland and is common is Tasmania. It is shrubby in age and
many feet high. The berries are yellow or orange-colored and usually
egg-shaped; they have proved poisonous to some animals when
taken in quantity. Dunal (conf. Cand. Prodr. xiii. Part i. 69) has
in following Nees of Esenbeck (Lehm. Plant. Preiss. i. 343) united

with S. aviculare an allied still specifically quite different West-Australian plant, S. fasciculatum (F. M. Fragm. Phyt. Austr. i. 123). In the colony of South Australia S. simile (F. M. in Transact. Phil. Soc. Vict. i. 19) takes its place. S. vescum (F. M. in Transact. Vict. Inst. 67 ; Plants of Vict. tab. lxii.) dubiously referred to S. aviculare by Dr. Hooker (Flor. Tasm. i. 288) is amply distinct and has recently been discovered also on the islands of Western Port. Whether the Timor plant (conf. Decaisne, Herb. Timor, 42) described by Dunal as S. aviculare in Candolle's Prodromus and previously introduced into his elaborate Synopsis Solanorum, p. 9, as S. glaberrimum, is identical with our southern species, requires to be further ascertained and seems improbable.

Mr. Travers alludes in his journal to a dwarf New Zealand Nightshade also found in the Chatham-group. This is unquestionably the now cosmopolitan S. nigrum, immigrated there like into Australia.

MYOPORINÆ.

MYOPORUM LÆTUM.

Forst. Prodrom. 238 ; All. Cunn. in Annals of Nat. Hist. i. 461 ; Alph. de Cand. in Cand. Prodr. xi. 709 ; J. Hook. Flor. Nov. Zeel. i. 204.

Common in the woods of Chatham-Island.

Though Mr. Travers's specimina are destitute of flowers and fruits, which were not obtainable at the time of his visit, there seems no doubt that the Chatham-plant is referable to the New Zealand species. The leaves are not quite so grossly dotted as generally those of the New Zealand plant, but like these often toothed far below the middle.

ASPERIFOLIÆ.

MYOSOTIDIUM NOBILE.

Hook. Bot. Magaz. t. 5137 ; Cynoglossum nobile, J. Hook. in Lindl. Gardn. Chronicle, 1858, p. 240.

Everywhere on the coast, but chiefly on sea-sand often moistened by salt-water.

According to Mr. Travers the natives eat the root and use the leaves for smoking. Pigs feed eagerly on the roots.

The following description is given chiefly from a brief essay on the vegetation of the Chatham-Islands read during the session of 1858 before the Philosophical Institute of Victoria, when a living flowering plant of this species was exhibited.

A perennial diffuse herb. Root thick, long, almost cylindrical. Leaves crowded at the neck of the rhizome, sometimes fully 1′ long, heart-shaped or the lowest sometimes renate, of a beautiful lustre, longitudinally seven-nerved, plicate, conspicuously veined, above dark-green and smooth, beneath paler and scantily strigulose, undulate at the margin, blunt or emarginate at the apex, at the base gradually tapering into the petiole, which is in its lower part channelled and ciliate and in its upper part almost cylindrical and but slightly furrowed. Stem-leaves few, roundish or oval, short-petioled or sessile. Flowers crowded in corymbose panicles, without scent. Peduncles ½″ long or shorter. Calyx deeply five-cleft, not enlarged in fructification, as well as the peduncles beset with short appressed grey hair ; its lobes orbicular-ovate, appressed, hardly longer than 1‴. Corolla short salver-shaped, with exception of the beards of the faux glabrous, five-lobed, at the junction of the lobes somewhat folded ; the lobes roundish, quincuncial in bud, 1¼–2½‴ broad, above of dark- beneath of pale-blue ; the tube measuring hardly 1‴, outside livid, inside yellowish-green ; fornices glabrous, opposite to the lobes of the calyx, measuring about ¼‴, outside cyan-blue, emarginate and pale at the apex, at the margin not distinctly papillate but finely wrinkled ; the faux between the fornices above the point of the insertion of the stamens slightly bearded. Stamens 5, enclosed, opposite to the lobes of the calyx. Anthers free, introrse, connivent, glabrous, yellow, ½‴ long, two-celled, dorsifixed, oblong-ovate, emarginate at the apex and base ; the cells bursting in their whole length. Pollen pale-yellow, consisting of globular grains. Filaments about as long as the anthers, glabrous, terete, inserted towards the summit of the corolla-tube. Ovaries smooth, innate with their base to the very narrow pale disk. Style cylindrical, about ¼‴ long or little longer, as well as the stigma pale and glabrous. Stigma bilobed, its divisions outward rounded. Carpidia about three times as long as the calyx, almost cordate, dorsally compressed, rather acute, glabrous, at the margin wing-like expanded and irregularly denticulated, 4–6‴ long, forming a tetragonous-hemispherical fruit, at their lateral acute margins free, protruding at their base. Seeds pendent from the summit of the cavity, oblique-ovate, turgid, smooth, about 3‴ long. Receptacle tetragonous-conical, as long as the carpels, terminated by the style. Cotyledons almost elliptical. Radicle very short.

One of the most singular and beautiful of the few endemic plants of the Chatham-Islands. Without flowers it resembles more a Funkia

E

than any coordinal species, and hence it may be that it passes under the strangely inappropriate name Chatham-Islands Lily. It seems to require absolutely a moist air, if not also a somewhat saline soil, to be maintained in cultivation.

As a genus Myosotidium ranks near Cynoglossum, to which it was simultaneously referred by Dr. Hooker and the author in his final notes communicated to the Victorian Institute. But as long as some allied plants are upheld in separate generic positions and thus Linné's original definition of Cynoglossum remains abandonéd, Myosotidium must be regarded, as the venerable Sir Will. Hooker well suggests, as of generic importance, a position to which moreover its strikingly peculiar habit gives it additional claims.

It seems that the leaves are attaining in the native locality still much larger dimensions than those recorded on this occasion. So ornamental a plant and one moreover not without its uses, might advantageously be transplanted to other shores, and might especially be naturalized on those of Australia and New Zealand.

PRIMULACEÆ.

SAMOLUS REPENS.

Persoon, Synops. Plantar. i. 171 (1805) ; S. litoralis, R. Brown, Prodrom. Fl. Nov. Holl. 428 (1810) ; Loddig. Cabin. t. 435 ; Duby in Cand. Prodrom. viii. 73 ; Nees in Lehm. Plant. Preiss. i. 337 ; Schlechtend. Linnæa, xx. 617 ; J. Hook. Flor. Nov. Zeel. i. 207 ; Flor. Tasm. i. 301 ; S. junceus, R. Br. l. c. 429 ; S. ambiguus, R. Br. l. c. ; S. campanuloides, R. Br. accord. to Ræm. & Schult. Syst. Veget. v. 2 ; Duby in Cand. Prodr. viii. 73 ; S. parviflorus, Nees in Lehm. Plant. Preiss. i. 337 ; Campanula porosa, Thunb. in Linn. fil. Suppl. Plant. 142 ; Prodr. Flor. Capens. 39 ; Sheffieldia repens, Forst. Charact. Gen. p. 18, t. 9 ; S. incana, Labill. Plant. Nov. Holl. Specim. i. 40, t. 54 ; Lysimachia sedoides, Lehm. Index Sem. Hort. Hamburg.

Stems erect, ascendent or prostrate, generally rigid, often perennial and woody towards the base, sometimes very tall ; root-fibres soon valid ; *leaves somewhat fleshy*, subulate- or lanceolate-linear or lanceolate, those of the stem occasionally those of the root always spatulate or obovate, the former sometimes reduced to minute linear-semilanceolate or deltoid very distant bract-like scales or even totally obliterated, not rarely as well as the stems and branches rough from prominent dots ; pedicels rather valid, at last from as long to twice as long as the calyx, seldom longer ; bracts lanceolate- and linear-subulate, often conspicuous and leaf-like ; flowers large ; *teeth of the flowering calyx semilanceolate or subulate-semilanceolate, longer than its tube;* corolla of firm consistence, half or more than half exserted; *filaments adnate up to the summit of the corolla-tube* and generally exceeding it in length ; *anthers large, almost ovate, cuspidulate; style and staminodia elongated; capsule large, firm, nearly ovate, about half emersed,* deeply valvate, the lobes of the calyx reaching to its summit or beyond it ; valves slightly divergent at the apex ; placenta conical, cuspidate ; seeds distinctly clathrate.

Near the sea, Chatham-Island. Found also in New Zealand, in South Africa, South America and around the whole extratropical coast of Australia and there also inland especially around brackish lakes and lagoons or on saline flats and even occasionally in mountain-districts (for instance on the cataracts of Mount Lofty).

The almost leafless varieties (S. junceus and S. ambiguus) are as yet only brought from S. W. Australia but pass there clearly into the normal form ; nor has the analysis of their flowers or fruits revealed the slightest difference from the typical plant. Analogous leafless varieties are produced under certain circumstances also by Tetratheca ericifolia, one of these being T. subaphylla (Benth. Flor. Austr. i. 132). S. campanuloides from South Africa is precisely in stature and external form alike to certain robust varieties observable in Australia, and shows also no differences in its floral or carpic characters.

One of the most variable of known plants, sometimes reduced to small tufts, sometimes fully 3′ high. Stem-leaves sessile or occasionally (particularly in the broad-leaved varieties) producing a distinct petiole, sometimes extremely minute or even transformed to scales, sometimes again to $1\frac{1}{2}''$ long and to $\frac{1}{4}''$ broad. Flowers terminating the branches as leafy or bracteate short or elongated racemes with several flowers or as few-flowered corymbs or even singly, or placed solitary axillary, the racemes not rarely constituting panicles ; the flowers sometimes very remote. Pedicels only exceptionally more than $1''$ long, usually shorter. Calyx and corolla five-lobed, rarely four- six- or seven-lobed. The latter cleft to half or two-thirds its length into semiovate lobes, glabrous, attaining a length of $5'''$, but almost always in various degrees shorter, but never so small as that of S. Valerandi. Staminodia as long as the stamens or somewhat longer or occasionally shorter, sometimes only at the apex free. Filaments in most cases extending considerably beyond the tube of the corolla, to the whole length of which they are adnate. Anthers $\frac{1}{2}$–$1'''$ long, with a more or less conspicuous terminal slender prolongation, introrse, with emarginate base. Style attaining a length of 1¼''', but generally shorter. Capsule 1½-3''' long, often conspicuously attenuated at the summit, with as many valves as divisions of the calyx, to which they are opposite. Seeds brown, plane-convex or compressed, roundish or somewhat angular, ¼-⅛''' long.

S. subnudicaulis (St. Hilaire accord. to Compt. rend. Acad. Scient. Paris, 1838, second sem. p. 98; Duby, l. c. viii. 74) agrees as far as the description permits of judging with S. litoralis.

Sprengel (Syst. Veg. i. 703) refers Androsace spathulata (Cavanill. Icon. v. 56, t. 484, f. 1; Samolus spathulatus, Duby in Cand. Prodr. viii. 74) also to this species, though the enclosed stamens point rather to S. Valerandi. Thunberg, according to Steudel (Nomenclat. Botanic. ii. 509) gave to S. litoralis or S. campanuloides the name S. porosus, perhaps nowhere distinctly published.

Persoon's name for the plant; S. repens (Synops. Plant. i. 171), in strict justice to priority ought to take precedence of that usually now adopted for the species, although it applies only to a variety; but neither is the name S. litoralis strictly applicable, inasmuch as the species occurs abundantly inland, and as the appellation was intended by R. Brown merely for the more ordinary not the almost leafless forms of the species. S. ebracteatus (Kunth in Humboldt & Bonpl. Nov. Gen. et Spec. ii. 223, t. 129; Rœm. & Schult. Syst. Veg. v. 2; Duby, l. c. viii. 74; Chapman, Flora of the Southern United States, 282) requires a renewed elucidation; its description accords neither with S. repens nor with S. Valerandi, though more with the former than with the latter; by the apparent want of staminodia it cannot be distinguished, since we may observe them in Australian specimina of S. repens occasionally also so short as not to extend beyond the portion adnate to the corolla-tube. Hence the depressed-globose capsule with inflexed valves and the enclosed stamens alone seem to distinguish it and if so probably not specificially. The short note on S. caulescens (Willd. accord. to Rœm. & Schult. l. c. v. 4) agrees equally with S. Valerandi and S. repens. S. Americanus (Spreng. Syst. Veg. i. 703) can from its brief and insignificant description likewise not be recognized. It appears thus not improbable that only two species of this genus exists, the Linnean one widely dispersed over the globe, the other restricted to the southern hemisphere, unless S. ebracteatus should be referable to our plant.

The occasion seems appropriate for furnishing a new diagnosis of the original species, although it has been found neither in New Zealand nor the Chatham-Islands.

SAMOLUS VALERANDI.

Linné, Spec. Plant. 243; Willd. Spec. Plant. i. 927; Desfontaines, Flor. Atlant. i. 183; R. Br. Prodr. 428; Smith, Prodr. Flor. Græc. i. 147; Smith, English Flora, i. 324; Rœm. & Schult. Syst. Vegetabil. v. 1; Duby in Cand. Prodr. viii. 73; Torrey, Flora of New York, ii. 13; S. aquaticus, Lam. Flor. Française. 329; S. floribundus, Kunth in Humb. & Bonpl. Nov. Gen. and Spec. ii. 181; Rœm. & Schult. l. c. v. 3; Duby, l. c. viii. 73; S. latifolius, Duby, l. c. viii. 74.

Smooth; stems erect, flaccid, herbaceous, seldom leafless; root-fibres tender; *leaves membranous,* lanceolate- or spatulate- or orbicular-obovate, scarcely ever passing gradually into bracts; pedicels capillary, several or many times longer than the calyx; bracts minute, linear- or lanceolate-subulate or lanceolate; flowers small; *teeth of the calyx deltoid, soon shorter than the tube; corolla tender; filaments surpassed in length by the corolla-tube, adnate to its base, thence free; anthers minute, cordate, not cuspidate; style and staminodia exceedingly short; capsule small, thin, globular, slightly emersed,* short-valved, the teeth of the calyx hardly reaching to or beyond its summit; valves strongly recurved at the apex; placenta spherical; seeds minute, almost smooth.

In Australia sparingly distributed from the Tambo and Snowy River eastward through Gipps-Land and northward through New South Wales and the southern parts of Queensland, occurring generally along forest-rivulets in rich soil.

The plant attains here a height of 2'. Leaves reaching a length of 2" exclusive of the petiole. Pedicels finally sometimes nearly 1" long, but often shorter. Corolla ⅔-1½''' long, beyond the middle five-cleft, bluntly lobed. Calyx at last about 1''' long, concealing almost the fruit. Filaments about as long as or little longer than the anthers; the latter only ⅛-¼''' long. Style sometimes almost obliterated. Capsule 1-1½''' long.

S. latifolius, according to a specimen gathered by Professor Philippi in Valdivia represents a variety with almost leafless stems and rather longer style.

The South African specimina of our collection are rather more robust than usual.

Dr. Dav. Dietrich (Synops. Plant. i. 715) draws as a synonym to this species Sedum alsinifolium (Allioni, Flor. Pedemont, 1740, t. 22, fig. 2), a plant referred by Candolle (Prodr. i. 404) to Sedum Cepæa.

MYRSINEÆ.

MYRSINE CHATHAMICA.

(Sect. Suttonia.)

Leaves obovate, toothless, pale-green on both pages, rather large, flat, on very short petioles ; calyx four-toothed, ciliolate ; pedicels twice or thrice shorter than the spherical drupes.

Common in the woods of the Chatham-Islands.

Branchlets almost glabrous. Leaves thinly coriaceous, 1–2″ long, immersed-dotted with roundish rufous glandular points, blunt or emarginate at the apex, finely net-veined, smooth, somewhat shining. Pedicels ¾–1½′″ long. Teeth of the deeply divided calyx about ½′″ long, semiovate-deltoid. Fruits (sent separately) obscurely purplish, one-seeded, of the size of large peas. Putamen faintly streaked. Seeds measuring about 2′″. Embryo and albumen normal.

The stature of the plant and the characters of its flowers remain to be recorded.

From its New Zealand congeners it is readily recognized ; thus Myrsine salicina differs already in the elongated form of its leaves and the comparatively long-stalked egg-shaped fruits ; M. Urvillei in undulated leaves of different color and smaller fruits ; M. divaricata in the smallness of its leaves and depressed drupes ; M. nummularia (Suttonia nummularia, J. Hook. Flor. Nov. Zeel. i. 173, pl. xlv.) also in the smallness of its leaves and smaller fruit.

Future investigations however must teach us, how far the adopted distinctive characters may be relied on, and whether not alpine localities and lowland-country will produce forms as different in the species of this genus as in Hymenanthera.

CONVOLVULACEÆ.

CALYSTEGIA SEPIUM.

R. Brown, Prodr. 483 ; A. Rich. Voy. de l'Astrolabe, i. 200 ; J. Hook. Flor. Nov. Zeel. i. 183 ; Fl. Tasman. i. 276 ; Torrey, Flora of New York, ii. 97 ; Miq. Flor. Indiæ Batav. ii. 624 ; Convolvulus sepium, Linné, Spec. 218.

Chatham-Island and Pitt-Island ; twining around shrubs.

The plant collected by Mr. Travers presents a transit from the normal form to C. tuguriorum, which species Dr. Hooker (Fl. Nov. Zeel. i. 183, t. 47 ; Convolvulus tuguriorum, Forst. Prodr. 74) upholds chiefly on account of its fulvous seeds. Achilles Richard unhesitatingly united the two plants (conf. Voy. de l'Astrolabe, i. 200), and his views were adopted by Choisy and by Raoul (Choix de, Plantes de la Nouvelle Zélande, p. 44). The author's observations on rather limited material tend to show that not only C. tuguriorum, but also C. silvatica, C. Soldanella, C. Dahurica and even C. spithamea are referable to C. sepium.

On the sandy beach of Gipps-Land the writer has observed the transit from the trailing plant with fleshy renate leaves to the typical form with hastate leaves and winding stems. New Zealand specimina of C. Soldanella or reniformis, communicated by Mr. Travers senior, have the ripe seeds black, smooth and about 3′″ long. If in reality these plants are distinct their marks of differences consist in others than those hitherto pointed out.

That C. sepium, when it adapts itself to sandy saline shores, should assume a prostrate growth and fleshy leaves of a somewhat altered form cannot be surprising, if we remember what an effect litoral localities exercise on the characters of other plants. Koch (Synops. Plant. Flor. German. et Helvet. ii. 568) inclines to refer C. silvatica also to C. sepium, a view fully borne out by the examination of Rochel's plant from the Banate in our collection. C. acutifolia, sent by Prof. Philippi from Valdivia, belongs also to C. sepium. C. Dahurica, according to a series of specimina, named by Turczaninow, flows likewise together with C. sepium and in its extreme forms links, as it seems, also to that widely diffused plant the North American C. spithamæa according to specimina of Dr. Short's and Dr. Sartwell's collections. To a certain extent analogous forms are produced by Clematis aristata in Australia. Calystegia affinis from Norfolk-Island is probably referable to C. marginata. The latter species (conf. J. Hook. Fl. Nov. Zeel. i. 184, t. 48) ranges from the mountain-glens in the vicinity of Port Phillip through S. E. Australia to the southern parts of Queensland. Its capsules measure hardly ¼″ and are spherical ; the seeds are brown-black, rough and about 1½′″ long. This species with many other Gipps-Land plants is likely yet to be found in the phytologically little explored north-east part of Tasmania. That some convolvulaceous plants are subject to extreme variations, generally proportionate to their adaptability to varied localities and climatic influences, is very evidently demonstrated by Convolvulus erubescens. Of the changes, which the external form of this plant in its almost universal range through extratropical Australia undergoes, an estimation may be formed when it is acknowledged, that the following plants have to be rejoined with it : C. angustissimus (R. Br. Prodr. 482), C. remotus

(R. Br. l. c. 483), C. acaulis (Choisy in Cand. Prodr. ix. 406), C.
Preissii, C. Huegelii, C. adscendens, C. subpinnatifidus (Vriese in
Lehm. Plant. Preiss. i. 346–347), C. crispifolius (F. M. in Linnæa,
xxv. 423).

GENTIANEÆ.

GENTIANA SAXOSA.

Forst. in Svensk Kongl. Vetenskap's Academien's Handinger, 1777, p. 183, t.
5; G. Forst. Prodrom. 132; Linné fil. Suppl. 175; Frœlich de Gentian.
23; Willd. Spec. Plant. i. 1273; G. Don, Gen. Syst. of Dichlam. Plant.
iv. 181; Rich. Voy. de l'Astrolabe, i. 202; Grisebach in Cand. Prodr. ix.
89; J. Hook. Flor. Nov. Zeel. i. 178; G. montana, G. Forst. Florul. Insular.
Austr. Prodrom. 133 (1786); R. Brown, Prodrom. 450; Rich. l. c. i. 203;
Grisebach, Gen. et Spec. Gentian. 236 & 362; J. Hook. Flor. Nov. Zeel.
i. 179; Fl. Tasm. i. 271; G. Diemensis, Griseb. Gent. 224; G. pleurogy-
noides, Gr. l. c. 236; G. Patagonica, Gr. l. c. 237; in Cand. Prodr. ix. 99;
J. Hook. Flor. Antarctic. 328, t. 115; G. Grisebachii, J. Hook. in Hook.
Icon. Plant. 636; G. bellidifolia, J. Hook. l. c. 635; G. concinna, J. Hook.
Fl. Antarctic. 53, t. 35; G. cerina, J. Hook. l. c. 54, t. 36; Pneumonanthe
saxosa, Schmidt in Rœm. Archiv. i. p. 10; P. montana, Schm. l. c.

Leaves verging into a roundish ovate lanceolate or linear-lanceolate form; flowers 1, 2 or few terminating the stems or branches, or
several or many forming an umbellate or corymbose or paniculate
cyme, usually provided with conspicuous pedicels; calyx quite herbaceous, half or more than half as long as the corolla, wingless, cleft
to three-fourths or half its length into equal semilanceolate or linear-
semilanceolate entire lobes; *corolla whitish, nearly bell-shaped,
neither appendiculate nor bearded nor fringed, very deeply cleft*
into entire ovate- or more oblong- or lanceolate-cuneate lobes, near
its base furnished with glandular nectar-cavities; filaments fringeless;
anthers free, soon versatile; stigmata distinct, subsessile; capsule
nearly cylindrical or ellipsoid-cylindrical, somewhat exserted, wingless, hardly stalked; seeds globular-ovate or spherical, smooth, wingless.

On fern- or grass-land or peaty soil of Chatham-Island.

This plant abounds on the higher mountains of New Zealand
and Tasmania, but seems to be not common in the warmer low-lands.
It is exceedingly abundant on meadows of the Australian Alps at an
elevation of 4000–6000', but rare on lower mountains or in lowland-
country, having only been found on Mount Macedon, at Port Fairy,
in the crater of Mount Gambier and in the Tattiara-country (on
Tilly's Swamp). It occurs further in Auckland- and Campbell-
Islands and in antarctic America. It is the only Australian species

of this elegant and useful genus, and with exception of G. prostrata also the only congener inhabiting the remote southern latitudes of America as far as is known, since also G. Magellanica (Gaudich. in Annal. des Scienc. Natur. v. 89) according to the description offered seems to exhibit only one of the many forms of G. saxosa.

Root simple or branched, variable in size, never very thick, frequently perennial or at least biennial, the withered stems of previous years' growth often visible on the flowering plant; in lowland localities the root perhaps annual, though seedlings flowering the first year may deceptively appear as annual plants. Stems erect or ascendent, from 1″ to 2′ high, singly or several or very many from the root, in the latter case the plant when richly in flower assuming a most ornamental appearance. Leaves flat, smooth, herbaceous or sometimes coriaceous, from 2‴ to 4″ long, 1–8‴ broad, of equal color and somewhat shining on both pages, thinly 3–5-nerved, the lateral nerves often partially obliterated; radical leaves tapering into a short or long petiole; the stem-leaves opposite, often sessile and distant, not vaginate; the floral leaves similar to the stem-leaves, but often smaller and occasionally alternate. Pedicels attaining exceptionally a length of 3″, but usually in various degrees shorter. Calyx 3–6‴ long, of nearly equal texture throughout; its lobes acute or acuminate, not reflexed, unconnected beyond the tube by any transparent membrane. Corolla quite smooth, membranous, persistent, fragrant, 5–12‴ long, generally to nearly three-fourths its length 5-cleft, occasionally 4- 6- or 7-cleft, of a waxy white and beautifully blue-veined, in some instances purplish- or pale-veined, the veins darker inside; its lobes acute or blunt; the tube yellowish. Filaments white or purplish, about half as long as the corolla, seldom shorter, one-nerved, linear, almost flat, attenuated at the apex and adnate base. Anthers blackish-green, turning yellow or bluish, ½–1‴ long; their cells narrow-ellipsoid, bursting on the inner side, unconnected below the middle. Pollen yellow. Glands between the lowest adnate portions of the stamens forming a slight cavity surrounded by a descendent somewhat callous almost semicircular or semielliptic line. Stigmata blunt, divergent or slightly revolute, almost semioval. Capsule subsessile, ½–1″ long, bursting towards the summit. Seeds finally dark-brown, shining, ½–¼‴ long.

The plant participates in the peculiar odor and bitterness of other Gentians.

It is evident that the number of described Gentianæ must be largely reduced.

F

EPACRIDEÆ.

Dracophyllum scoparium.

J. Hook. Flor. Antarctic. i. 47, t. 33 ; Flor. Nov. Zeeland. i. 170.

Common in the woods of Chatham- and Pitt-Island ; found also in New Zealand and Campbell-Island. ·

This plant attains, according to Mr. Travers's notes, sometimes a height of 40′. In Australia and Tasmania solely Richea pandani-folia, Cystanthe procera, Monotoca elliptica (a genus embracing only two or perhaps even only one genuine species) and exceptionally Trochocarpa laurina and Epacris heteronema rival it in height. In New Zealand possibly other Dracophylla may under favorable circumstances attain similar dimensions, as several are showing a tendency to an arborescent growth.

Leaves of the young-plant attaining a length of nearly 1′ and a width of almost 1″; these are very gradually attenuated and thus resemble those of D. verticillatum ; as observed by Mr. Travers they become shorter with the growth of the tree ; the leaves of the flowering branches inclusive of their clasping base ¾–3″ long. Spikes containing not rarely as many as eight flowers. Pedicels often glabrous. Clasping part of bracteæ with exception of the margin smooth. Sepals ovate- or narrow-lanceolate, about as long as the corolla-tube, gradually pointed. Corolla 3–4‴ long ; its lobes half or nearly as long as the tube, deltoid, acuminate. Style ½–1½‴ long. Capsule measuring nearly 1¼‴, globose, with depressed vertex.

The number of the species of this genus is evidently over-estimated.

An undescribed species of Dracophyllum with the habit of Richea scoparia, very distinct from D. verticillatum (D. montanum, R. Br.), occurs amongst the many new plants discovered by Mons. Pancher in New Caledonia and will be elucidated by Mons. Adolphe Brogniart in his forthcoming work on the vegetation of that island with other Epacrideæ as yet new to science.

Cyathodes acerosa.

R. Brown, Prodrom. Flor. Nov. Holland. 540 ; Rœm. et Schult. Syst. Veget. iv. 473 ; All. Cunn. in Annals of Nat. Hist. ii. 47 ; Don, Gen. Syst. iii. 776 ; Cand. Prodr. vii. 741 ; J. Hook. Flor. Nov. Zeel. i. 163 ; C. Oxycedrus, R. Br. l. c. 540 ; J. Hook. Flor. Nov. Zeel. i. 164 ; Flor. Tasm. i. 246 ; C. parvifolia, R. Br. l. c. 540 ; J. Hook. Fl. Nov. Zeel. i. 164 ; Flor. Tasm. i. 246 ; C. divaricata, J. Hook. Fl. Tasm. i. 246, tab. 74 B ; Epacris arborea, Murray, Syst. Veget. 198 ; E. juniperina, Forst. Charact. Gener.

20, t. 10, fig. n.; Forst. Prodr. 71; Ardisia acerosa, Gært. de Fruct. et
Semin. ii. 78, t. 94, fig. 2; Leucopogon Forsteri, Ach. Rich. Voy. de
l'Astrolabe, i. 216; Styphelia Oxycedrus, Labill. Nov. Holl. Plant. Specim.
i. 49, t. 69; Lissanthe acerosa, L. Oxycedrus, L. parvifolia, Spreng. Syst.
Veget. i. 660; L. divaricata, J. Hook. in Lond. Journ. of Bot. vi. 267.

Leaves scattered, spreading, distinctly petioled, subulate- or
lanceolate-linear, rarely oblong- or oval-lanceolate, mucronate, seldom
blunt, beneath pale grey and lined with 3–5 or more thin nerves,
above not distinctly streaked ; flowers solitary, on densely bracteate
pedicels ; sepals blunt, ciliolated, as long as the tube of the corolla
or sometimes considerably shorter ; lobes of the corolla ridged, as
well as the faux glabrous or sometimes slightly hairy ; filaments
often almost entirely adnate ; style very short ; drupe depressed-
globose, red ; pericarp thick ; putamen five-celled or occasionally
with fewer cells.

Varietas latifolia, J. Hook. Fl. Nov. Zeel. i. 163.
Chatham-Island. Capt. Anderson.

The broad- and long-leaved variety appears to be known only
from Chatham-Island, from whence it was first brought by Dr.
Dieffenbach. It differs from all the forms of C. acerosa collected in
New Zealand, Australia and Tasmania in larger often not mucronate or even quite blunt leaves, which attain a length of 1″ and a
width of 2 or 3‴ and are then 7–11-nerved. The drupes are also
unusually large, measuring nearly ¼″ in diameter. The corolla of
the Chathamian plant is 1¼‴ long, glabrous; its tube equalling in
length the sepals ; the lobes are rather semilanceolate, whilst in
Tasmanian specimina they are often deltoid ; they are either glabrous
or slightly bearded by hardly perceptible downs. Anthers oval or
ellipsoid.

The examination of a rich series of plants of our collection has
led to the union of all the plants enumerated here under C. acerosa.
The variability of the leaves is equally great in Monotoca or still
greater ; the length of the tube of the corolla affords no reliable
note of discrimination, as seen on forms of this species gathered on
the granite-rocks of Wilson's Promontory ; the scantily hairy faux
and lobes, which generally characterize together with the longer
corolla-tube C. divaricata, are also of no avail for distinction, since the
short-flowered variety occurs at Wilson's Promontory with equally
bearded corolla. Moreover Lissanthe montana (R. Br. Prodr. 540)
and Leucopogon Hookeri (Sond. in Linnæa, xxvi. 248 ; J. Hook. Fl.
Tasm. i. 251, tab. 75 B ; L. obtusatus, J. Hook. in Lond. Journ. vi.

269) are also only varieties of one species with either glabrous or bearded flowers. The leaves of C. acerosa are unusually large in plants produced by the mild climate of Chatham-Island, and very small in the alpine variety (C. parvifolia); the veins divergent from the more exterior nerves of the leaves may also be seen in Australian specimina whenever the leaves are widened towards the summit. On a former occasion (conf. Frag. Phyt. Austr. iv. 105) attention has been drawn to the fallibility of the characters on which the diagnosis of Leucopogon juniperinus and L. Fraseri rest; the latter indeed seems merely an alpine state of the former with a more densely bearded limb of the corolla; but hitherto only this alpine variety of L. juniperinus and not its taller normal form has been known from New Zealand. Cyathodes abietina shows the degree of pubescence of the corolla to be equally variable.

Cyathodes acerosa occurs in Tasmania as well on the coast as on alpine summits. In Australia it seems exclusively litoral and extends not, as far as known, west of Phillip Island. At Sealer's Cove on granitic declivities occasionally washed by the spray of the sea the plant rises exceptionally to a height of 20' and yields then a stem of ¾' thickness. When loaded with its carmin-red or pale rose-colored drupes it forms a most showy object in the surrounding vegetation.

Nearest in affinity and appearance to C. acerosa are C. glauca and C. abietina; the latter, which forms a shrub 1-2' high on rocks exposed to the sea on the south-coast of Tasmania, differs from the large-leaved variety of C. acerosa in less patent more rigid leaves with thicker nerves, of a pale-green but not glaucous color beneath; moreover the corolla is generally much larger and often densely bearded, whilst the lobes show no distinctly elevated axis. C. glauca is still more distinct by its leaves, which are crowded in intervals, above perceptibly streaked, less distinctly petioled and of a different nervature of numerous close and fine lines and less pale color beneath, by its sessile flowers, usually evidently bearded corolla of larger size, somewhat exserted filaments, longer anthers and style and often 8-10-celled fruits.

Cyathodes straminea and C. adscendens occur both on the western mountains of Tasmania; the latter species is transferable to Leucopogon, if the position of that genus can be vindicated.

Cyathodes empetrifolia (J. Hook. Fl. Nov. Zeel. i. 164; Androstoma empetrifolia, J. Hook. Fl. Antarctic. p. 44, t. 30) produces not rarely 3- or 4-flowered spikes.

Leucopogon Richei.

R. Brown, Prodrom. Flor. Nov. Holl. 541; Rœm. et Schult. Syst. Veg. iv.
475; Hook. Bot. Mag. 3251; J. Hook. Flor. Tasm. i. 249; F. M. Fragm.
Phyt. Austr. iv. 123; L. polystachyus, Loddig. Cabin. t. 1436; L. apicu-
latus, Sm. in Rees's Cyclop.; L. parviflorus, Lindl. in Bot. Reg. t. 1560;
Sond. in Lehm. Plant. Preiss. i. 305; Styphelia Richei, Labill. Nov. Holl.
Plant. Specim. i. 44, t. 60; Poiret, Encyclop. Méthod. vii. 433; S. parviflora, Andr. Reposit. t. 287; S. Gnidium, Vent. Malmais. t. 23.

On sandy ground near the sea abundant in Chatham-Island,
rare in Pitt-Island.

Mr. Travers found this Leucopogon in flower, and so far the
Chatham-plant accords precisely with the Australian species. The
fruit however still remains to be compared.

This is one of the few plants common to Australia and the
Chatham-group, yet not known as existing in New Zealand.

SCROPHULARINÆ.

Veronica Forsteri.

V. salicifolia, Forst. Prodrom. 11; Vahl, Symbol. Botanic. iii. 4; Vahl, Enum.
Plant. i. 67; Rœm. et Shult. Syst. Veget. i. 103; Spreng. Syst. Veget. i.
74; Ach. Rich. Voy. de l'Astrol. i. 186; A. Cunn. in Annal. of Nat. Hist.
i. 457; Endl. Annalen des Wiener Museums, i. t. 14; Lindl. Bot. Reg.
32, t. 5; Benth. in Cand. Prodrom. x. 459; J. Hook. Flor. Nov. Zeel. i.
191; V. elliptica, Forst. Prodr. n. 10; J. Hook. Fl. Ant. i. 58; Lindl. &
Paxton, Flower Gard. iii. 101; V. decussata, Soland. in Ait. Hort. Kew.
first edit. i. 20; Bot. Magaz. t. 242; Hombron & Jacquinot, Voy. au Pôle
Sud, t. 9; V. macrocarpa, Vahl, Symbol. iii. 4; V. parviflora, Vahl, l. c.;
V. angustifolia, A. Rich. l. c. 187; V. ligustrifolia, Hook. Bot. Magaz.
3461; V. speciosa, Rich. Cunn. in Bot. Mag. 3461 & 4075; Paxton, Magaz.
of Bot. x. 247; Houtte, Flor. des Serres, i. 17; iii. 197; V. diosmifolia,
Rich. Cunningh. in Bot. Mag. 3461; V. Lindleyana, Paxt. Magaz. xii.
247; Flor. des Serres, ii. 17; V. odora, J. Hook. Fl. Antarct. i. 62, t. 41;
V. stricta, Banks & Soland. in Cand. Prodrom. x. 459; V. myrtifolia,
Banks & Sol. l. c.; V. macroura, J. Hook. in Cand. Prodrom. x. 459; V.
Dieffenbachii, Benth. in Cand. Prodrom. x. 459; V. acutiflora, Benth. l. c.;
V. Menziesii, Benth. l. c.; V. lævis, Benth. l. c.; V. buxifolia, Benth. l. c.;
V. Andersonii, Lindl. & Paxt. Flower Gard. ii. t. 38; Houtte, Flor. des
Serres, vii. 568; V. Fonkii, Phillipi in Linnæa; Hebe Magellanica, Juss.
in Gmelin. Syst. Veget. p. 27.

Shrubby or arborescent, evergreen, erect; *leaves* opposite, char-
taceous or coriaceous, glabrous or very finely ciliated, *sessile or on
extremely short petioles, roundish or oval or broad- or narrow-*

lanceolate or linear-lanceolate, generally entire, seldom minutely and distantly serrulated, always one-nerved and very faintly veined ; *racemes densely many-flowered, less frequently few-flowered*, axillary or occasionally some terminal; peduncles generally short, seldom wanting, mostly as well as the rachis pedicels and calyces subtle-downy; pedicels about double as long as the calyx or variously shorter; corolla bilabiate ; *capsule ovate or orbicular-ovate, moderately or slightly parallel to the septum compressed, by septical dehiscence divisible into two carpels*, often longer than the calyx ; carpels bursting at the summit and with a narrow slit along their internal face ; placenta stalked ; seeds flat, fulvous, orbicular or broad-ovate, smooth.

Varietas salicifolia.

Common in woods of Chatham-Island and around their edges.

This variety embraces those more luxuriant forms of V. Forsteri, in which the leaves and racemes are elongated, and in which not unfrequently the tube of the corollæ considerably extends beyond the calyx, whilst its lateral and upper lobes are oval or oblong and stamens and style long-exserted. This variety is also frequent in New Zealand, but seems as yet unknown from any other localities except the Chonos-Archipelago, where it occurs according to Professor Philippi's collection.

Varietas elliptica.

Common on Pitt-Island, rare on Chatham-Island.

This variety comprises those forms of V. Forsteri, which are much reduced in the length of their leaves and often simultaneously also in the length of their racemes ; the pedicels and tube of the corolla are also often shorter than in the var. salicifolia and the lobes of the corolla broader, whilst stamens and styles are generally less elongated.

This variety inhabits not only New Zealand, but also Auckland-and Campbell-Island, Fuegia and the Falkland-Islands.

From the rich collections of New Zealand plants, which the phytological museum of Melbourne owes to Mr. Travers senior and Dr. Haast, it appears that this variety is mainly restricted to rocky and exposed coast-localities and alpine regions. Both varieties pass under the native name " Koromiko" at the Chatham-Islands.

After instituting extensive dissections the author has been induced to unite all the plants enumerated on this occasion as forms of one species, and it seems not improbable that also V. pubescens (Banks & Soland. in Cand. Prodr. x. 460) and V. pimeloides (J. Hook.

Fl. Nov. Zeel. i. 195) belong to the array of forms, for which, to avoid misunderstanding, the collective name V. Forsteri has been now adopted. The forms which it comprises, though in their extremes habitually so dissimilar, are clearly linked together by an uninterrupted chain of graduations. Other plants either of wide distribution or accommodating themselves to very different conditions of climate or soil are presenting if not analogous at least quite as singular aberrations of forms as may be noticed in V. Forsteri ; as such we may instance Bursaria spinosa, Dodonæa viscosa, Tetratheca ericifolia, Epilobium tetragonum, Vittadinia cuneifolia, Correa speciosa, Beyera viscosa, Convolvulus erubescens, Samolus repens. In some of the glacier-forms of V. Forsteri the bracteoles are enlarged, the capsules shortened or the calyces lengthened, and the inflorescence may become truly spicate or even capitate. In by no means rare instances plants with the small leaves of V. elliptica produce the long racemes of V. salicifolia. In culture the varieties are more constant than usual ; thus the handsome narrow-leaved form, so long acknowledged as a species under the name V. parviflora, has undergone no change for a series of years in our garden. But the collections of dried plants received from New Zealand exhibit unequivocal transits as well to V. salicifolia as to V. elliptica. What has been effected in changing the color of the flowers of this plant by artificial fecundation of its varieties may also to some extent be effected in Kennedya monophylla.

URTICEÆ.

URTICA INCISA.

Poiret, Tableau Encyclopédique, Suppl. 223 ; Weddell in Archives du Muséum d'Hist. Nat. Paris, ix. 81 ; J. Hook. Flor. Tasm. i. 343 ; U. lucifuga, J. Hook. in Lond. Journ. of Bot. iv. 285 ; Flor. Nov. Zeel. i. 225.

On the edges of bushy land of Chatham-Island.

The plant brought by Mr. Travers is broad-leaved and monœcious. The narrow-leaved form occurs in New Zealand, Tasmania and scantily in Australia felix ; the broad-leaved variety is known from the forest-country of East Gipps-Land and various parts of East Australia as far north at least as the southern boundaries of Queensland.

The differences, if really specific, between this plant and the protean U. dioica, which in our collections occurs from many distant parts of the globe, have yet to be further traced.

The diagnostic circumscriptions of the other described Nettles of this part of the globe (U. ferox, U. australis, U. Aucklandica) need also a renewed revision. U. urens and U. dioica are now widely disseminated through Australia.

In the forests of tropical East Australia occurs a probably distinct species of this genus.

PIPERACEÆ.

MACROPIPER EXCELSUM.

Miquel, Systema Piperacear. 221; Piper excelsum, Forst. Prodr. 20; Vahl, Enum. Plant. i. 335; Kerner, Hort. Sempervir. 33; Rœm. & Schult. Syst. Veg. i. 313; All. Cunn. in Annal. of Nat. Hist. i. 210; J. Hook. Fl. Nov. Zeel. i. 227.

In woods of Chatham-Island not common.

Mr. Travers found the plant neither in flower nor fruit; but there seems no doubt of its identity. The Chathamians bestow like the New Zealanders on it the appellation "Kawa Kawa." If, as Dr. Hooker's researches tend to prove, the same species extends to Norfolk-Island, it is then evident that also Macropiper psittacorum (Miq. Syst. Piper. 221; Piper psittacorum, Endl. Prodr. Flor. Insul. Norfolk, 37) are referable to M. excelsum.

THYMELÆÆ.

PIMELEA ARENARIA.

All. Cunningham in Botan. Magaz. 3270; Annals of Natur. Histor. i. 377; J. Hook. Flor. Nov. Zeel. i. 221; Meisner in Cand. Prodrom. xiv. 517; P. sericeæ var. Wikstrœm in Actis Academ. Scient. Holmens. 1818, 282; Passerina villosa, Thunberg, Mus. Nat. Acad. Upsaliens. xiii. 106; Gymnococca arenaria, C. A. Meyer, Index Sem. Hort. Petropol. 1845; Bulletin Acad. St. Petersb. iv. 171; Annal. des Scienc. Naturell. trois. série, v. 373; Endl. Gen. Plant. Suppl. iv. p. ii. 60; Walpers, Annal. Botan. System. i. 586.

On rocky places of Chatham-Island, where it was also found by Dr. Dieffenbach.

The capitula contain not unfrequently from 14–17 flowers.

Our collections afford not sufficient material for drawing anew the real specific limit of the plant, of which the silver-silky P. arenaria appears to represent merely a certain state. Had the opportunity been afforded by our collection of examining the fruits of all hitherto described New Zealand species of this genus, and had their carpological characters proved identical, all the following plants would probably have been united as varieties of one species, variable according to soil, elevation of locality and other circumstances, as multifarious transitory forms in the collections of Mr. Travers and Dr. Haast at least to some extent appear to prove : Pimelea Gnidia, Banks & SoL accord. to Meisn. in Candolle Prodr. xiv. 517 ; P. prostrata, B. & S. l. c.; P. pilosa, Willd. Spec. Plant. i. 50 ; P. virgata, Vahl, Enum. Plant. i. 306 ; P. Urvilleana, A. Richard, Voy. de l'Astrolabe, i. 227 ; P. arenaria, All. Cunn. l. c.; P. Lyallii, J. Hook. Fl. Nov. Zeel. i. 223 and the plants already reduced to those by Dr. Hooker and Prof. Meisner.

As pointed out by All. Cunningham P. sericea bears a close resemblance to P. arenaria ; but scarcely less so to P. ammocharis (F. M. in Hook. Kew Miscell. 1857, 24 ; and in Edinburgh New Philos. Journ. 1863, 233). The latter however has diclinous flowers and the leaves on both pages silky ; the former has stamens and style more distinctly exserted, and both species have seemingly not the succulent fruit ascribed to P. arenaria.

The number of described species of Pimelea is far in excess of that limited by nature, and much has yet to be learned on their distribution ; thus Pimelea axiflora, a species ascending from the forest-valleys of the lowlands to the glacier-regions, has recently been found in the southern parts of New South Wales and also in King's Island, a dependency of Tasmania. P. microcephala again has been collected by Oldfield on the Murchison-River, to which locality many plants of the Murray-desert are extending.

POLYGONEÆ.

POLYGONUM MINUS.

Hudson, Flor. Anglic. first edit. 148 ; Smith, Engl. Flora, ii. 235 ; Koch, Synops. Flor. German. ii. 712 ; Meisn. in Lehm. Plant. Preiss. i. 623 ; Meisn. in Cand. Prodrom. xiv. 111 ; J. Hook. Flor. Tasman. i. 306 ; P. Persicariæ var. Linn. Spec. Plant. i. 518 ; P. strictum, Allioni, Flor. Pedemont. 2051, t. 68, f. 2 ; Meisn. Monograph. Gen. Polygon. 74 ; in Wallich. Plant. Asiat. Rarior. iii. 57 ; Wight, Icon. Plant. Ind. Orient. v. tab. 1800 ; P. decipiens, R. Br. Prodrom. 420 ; P. prostratum, Ach. Rich. Voy. de l'Astrolabe, i. 177 ; J. Hook. Flor. Nov. Zeel. i. 209.

In swampy places of Chatham-Island.

The plant gathered by Mr. Travers is in incipient inflorescence and so far agrees with the definition of P. minus. The species is widely distributed through extratropical Australia. Notes on the range of several Australian species of this genus are contained in the

G

report on the botanical results of A. Gregory's expedition (conf. Proceedings of the Linnean Soc. 1858, 149–152).

MUEHLENBECKIA AUSTRALIS.

Meisner, Gener. Plantar. Vascul. p. 316; Meisn. in Cand. Prodr. xiv. 146 ; Coccoloba australis, Forst. Prodrom. 176 ; Polygonum australe, A. Rich. Voy. de l'Astrolabe, i. 178; Endlich. Prodrom. Flor. Insul. Norfolk. 42 ; Endl. Iconograph. Gener. t. 87 ; J. Hook. Flor. Nov. Zeel. i. 210; P. australe and P. appressum, All. Cunn. in Annal. of Natur. Histor. i. 455 ; Raoul, Choix de Plantes de la Nouvell. Zélande, 43.

On edges of woods of Chatham-Island.

The plant presents the normal form. It is a tall climber ; the leaves are not rarely three-lobed and vary considerably in size.

The limits between M. australis, M. appressa, M. axillaris (M. parvifolia, Meisn. in Linnæa, 1853, 362) and M. complexa have to be traced anew. In Australia felix M. appressa occurs always as a lowlands plant, whilst M. axillaris is restricted to subalpine mountains. It seems evident from plants collected in New Zealand by Mr. Travers senior and Dr. Haast, that M. complexa exhibits merely a luxuriant state of M. axillaris.

In the Fragmenta Phytographiæ Australiæ (vol. iv. 131) it has been pointed out why at least the Australian species of Muehlenbeckia might well be transferred to Polygonum.

ORCHIDEÆ.
EARINA MUCRONATA.

Lindl. Bot. Regist. 1699 ; All. Cunn. in Hook. Companion to the Bot. Mag. ii. 377 ; Hook. Icon. Plant. t. 431; J. Hook. Fl. Nov. Zeel. i. 239; Epidendrum autumnale, Forst. Prodrom. 319 ; Cymbidium autumnale, Swartz in Nov. Act. Soc. Reg. Upsal. vi. 72 ; Willd. Spec. Plant. iv. 28; A. Rich. Voy. de l'Astrolabe, i. 169.

Chatham-Island, on stems of fern-trees.

Mr. Travers's plant is fruit-bearing. The capsules are about ⅓" long, narrow-ellipsoid and strongly ribbed.

PTEROSTYLIS BANKSII.

R. Brown, accord. to All. Cunn. in Bot. Mag. t. 3172; All. Cunn. in Hook. Compan. to the Bot. Mag. ii. 376 ; Lindl. Gener. et Species Orchid. 388 ; J. Hook. Fl. Nov. Zeel. i. 248.

On grassy places of Chatham-Island.

The plants of Mr. Travers's collection are unusually dwarf, some only of a finger's length.

Varietas silvicultrix.

Chatham-Island, in woods only.

The characters of this variety consist in broader and shorter leaves, which are verging from broad-ovate into lanceolate, only 1-2½" long, but ⅜-1" broad and acute but not acuminate, in proportionately broader sepals, of which the inner are lanceolate and simply acute, whilst the outer are hardly or little longer than these and never so much protracted into a narrow acumen as those of the typical form of Pterostylis Banksii. The author however has been unable to detect any important structural differences between these plants and has therefore not ventured to separate them as species, although middle-forms are missing in the collection. New Zealand specimens of P. Banksii prove that plant subject to considerable changes in its external form.

CHILOGLOTTIS TRAVERSII.

Caladenia bifolia, J. Hook. Fl. Nov. Zeel. i. 247.

Leaves subsessile, lanceolate or ovate-lanceolate, *as well as the scape bract and pedicel short- and glandular-downy ;* sepals conspersed with glands ; upper one broad-cymbiform, simply acute ; lower ones linear-lanceolate, almost of equal form with the lateral somewhat smaller sepals ; *labellum only along the mid-line and below the middle glandular,* sessile, obovate, blunt, not decidedly appendiculate ; column near the almost lobeless summit somewhat dilated.

Amongst ferns as well in Pitt-Island as in Chatham-Island.

Leaves 1-2" long. Scape measuring 3-5". Pedicel during anthesis about half exserted beyond the clasping bract. Upper sepal and the two lower slightly longer than the lip ; the latter about 4‴ long ; its glands biseriate, not very large, increasing downward in size.

The only New Zealand specimen, which the author had an opportunity of dissecting, and which was collected in the Province of Canterbury, showed no differences in the organization of its flowers from those of the Chatham-plant, but its leaves are broad-ovate.

Dr. Hooker (Flor. Tasm. ii. 23) alludes to the exclusive existence of a Chiloglottis in Lord Auckland's Group and Campbell-Island, but not to the New Zealand plant, as a Caladenia doubtfully combined with an Auckland species in his Flora Nov. Zeel.

Chiloglottis Gunnii (Lindl. Gen. et Spec. Orchid. 387 ; J. Hook. Fl. Tasm. ii. 23, t. 108 B.) occurs on the mossy stems of fern-trees in various localities of the colony of Victoria. The following diagnosis and description resulted from the examination of the living plant.

Leaves ovate or oblong or lanceolate, glabrous, distinctly petioled; *upper sepal ovate-cymbiform*, acuminate, about twice as broad as the lateral subfalcate-lanceolate sepals; *labellum sessile, cordate-ovate, only near the axis glandular-tuberculate;* apex of the column bidentate.

Tuber globose-ovate or exactly ovate. Leaves 1–2″ long, beneath paler. Petioles generally from ½–1″ long, channelled. Peduncle 1″ long or longer, finally sometimes much lengthened, cylindrical. Bracteole acuminate, about ½″ long. Sepals all green ; the lower from a broader base linear, long and gradually attenuated, channelled, hardly however on the very summit terete, slightly longer than the lip, about as long as and much narrower than the lateral sepals, almost horizontal and lightly recurved at the apex ; upper sepal nearly 4‴ broad, sometimes faintly tinged with purple. Labellum hardly longer than the column, towards the apex almost flat and somewhat purplish, towards the blunt greenish basis lightly inflexed ; its two basal glands half-adnate, purple ; the terminal gland sessile, roundish heart-shaped ; the two next ones oblong; the rest on a short or very short stipes, brown-purple. Column pale-green, spatulate, slightly curved.

The differences, expressed by these notes from Mr. Archer's beautiful illustration of the Tasmanian plant, seem not of specific value ; but the Victorian plant may as a variety perhaps be distinguished as viridiflora. Dr. Jos. Milligan found C. Gunnii on St. Mary's plains, the Hampshire hills, and on the alpine summit of Ben Lomond ; it occurs also at Southport.

Chiloglottis diphylla ranges in East Australia as far north as Moreton Bay.

It is not improbable that still other terrestrial Orchids will be found existing in the Chatham-group.

IRIDEÆ.

LIBERTIA IXIOIDES.

Sprengel, Syst. Veg. i. 168 ; Reichenbach, Iconograph. Botan. Exotic. t. 157 ;
J. Hook. Flor. Nov. Zeel. i. 252 ; L. grandiflora, Sweet, Hort. Brit. 498 ;
Sisyrinchium ixioides, Forst. Prodrom. 325 ; Ach. Rich. Voy. de l'Astro-
labe, i. 161 ; Ferraria ixioides, Willd. Spec. Plant. iii. 582 ; Moræa
ixioides, Thumb. Dissert. de Moræa, p. 8 ; Renealmia grandiflora, R. Br.
Prodrom. Flor. Nov. Holl. 592; Sweet, Flower-Gard. t. 64; Loddig. Cabin.
t. 993 ; Nematostigma ixioides, Alb. Dietr. Syst. Willd. 228 ; Dav. Dietr.
Synops. Plant. i. 150.

Chatham-Islands, on open places amongst grass.

The scape is sometimes reduced to extreme shortness and thus long surpassed in length by the leaves.

The geographical distribution of the species of this genus is interesting. L. ixioides occurs widely dispersed in New Zealand and as remarked in Chatham-Island. L. micrantha seems restricted to New Zealand. L. paniculata verges northward in Australia to Mount Lindsay of Queensland, where it was collected by Mr. Walt. Hill, and southward to the entrance of the Snowy River in Gipps-Land, and may hence be sought for in the N. E. extremity of Tasmania. L. pulchella is known only from the vicinity of Port Jackson, and specimina kindly communicated by Mr. W. Woolls, the assiduous examiner of the vegetation around Parramatta, show this species to produce black seeds, whilst those of the five other Australian and New Zealandian species are brown. L. Lawrencii is not unfrequent in the humid especially higher forest-regions of Tasmania, but in the Australian Continent only found on the summit of Mount Juliet, and on springy subalpine localities at the remotest sources of the Yarra and La Trobe-River and their tributaries, usually consociated with Oxalis Magellanica. L. azurea (L. stricta, L. laxa, L. graminea, Endl. in Lehm. Plant. Preiss. ii. 32 ; Orthrosanthus multiflorus, Sweet, Flor. Austral. 11, t. 11 ; Loddig. Cabinet, 1474 ; Paxton Magaz. xi. 245 ; Wuerthia elegans, Regel, Garten-Flora, ii. 46) extends in the vicinity of the coast-line from Cape Nelson of Portland Bay to West Australia, being gathered in very many intermediate spots of the mainland and also on Kangaroo-Island. The latter plant is remarkable for the want of pedicels, the flowers being sessile and clustered within the bracts and bracteoles. Its beautiful blue color distinguishes it also from all congeners except the Chilian L. cœrulescens (Kunth in Linnæa, xix. 382). L. formosa (Graham in Lindl.

Bot. Registr. 1630, Hook. Botan. Magaz. 3294) of Chiloe seems the only other species known but those mentioned on this occasion.

LILIACEÆ.

PHORMIUM TENAX.

J. R. and G. Forster, Charact. Gen. 24; Linn. fil. Suppl. Plant. 204; Cook's Voyage, ii. 96, with Illustr. ; Mill. Fascic. tab. 1; Thunberg, Nov. Plant. Gener. p. 94; Redouté, Liliac. t. 448 & 449; Desvaux, Journ. iv. tab. 17 & 18; Thouin in Annal. du Mus. Paris, ii. 224 & 474; St. Fond in Annal. du Mus. xix. 40, t. 20; Thibaud in Annal. Soc. Linn. Paris, iv. 57, t. 7; Diction. Scienc. Natur. t. 51; Herbier Amat. t. 120; Kerner, Gen. t. 116 & 117; Ach. Rich. Voy. de l'Astrolabe, i. 153; Bot. Magaz. 3199; Endl. Prodrom. Flor. Insul. Norfolk. 27; Kunth, Enumer. Plant. iv. 275; J. Hook. Flor. Nov. Zeel. i. 156; P. Cookianum, Le Jolis, Mém. sur. le Lin de la Nouv. Zél.; Chalmydea tenacissima, Banks & Soland. in Gærtn. de Fruct. & Semin. i. 71, t. 18; Lachenalia ramosa, Lamark, Encyclop. iii. 373.

Recorded by Dr. Dieffenbach and all subsequent visitors as indigenous to the Chatham-group.

ASTELIA BANKSII.

All. Cunn. in Hook. Compan. to the Bot. Magaz. ii. 374; J. Hook. Flor. Nov. Zeel. i. 260; A. Richardi, Endlich. accord. to Kunth Enumer. iii. 365; Hamelinia veratroides, A. Rich. Voy. de l'Astrolabe, i. 158, t. 24.

On open places of Chatham-Island, not common.

Mr. Travers's collection contains the female plant in fruit.

Steudel unites with our plant (Synops. Plantar. Glumac. ii. 312) an allied but evidently distinct plant of the Sandwich-Islands, A. veratroides, Gaudich. in Freycen. Voy. de l'Uranie 420, t. 31.

SMILACEÆ.

RHIPOGONUM SCANDENS.

Forst. Charact. Gener. 50, t. 25; Edit. German. Kerner. 51, t. 25; Lamark, Enclyclop. t. 254; A. Rich. Voy. de l'Astrolabe, i. 151; J. Hook. Flor. Nov. Zeel. i. 253; R. parviflorum, R. Br. Prodr. 293; Kunth, Enum. Pl. v. 271; Smilax Rhipogonum, Forst. Prodrom. 372.

Common in woods of Chatham-Island. "Hartwhan" of the natives; "Supplejack" of the New Zealand colonist.

Of this genus three other species are known, all East-Australian, the R. dubium (Endl. Flor. Insul. Norfolk. 30) being a Clematis.

R. album (R. Br. Prodr. 293) extends from East Gipps-Land to
the tropic of Capricorn through the forest of East Australia. Two
other species or perhaps varieties are likewise found in subtropical
East Australia, which need further examination on the spot for the
purpose of ascertaining their relative position to R. album. (Conf.
F. M. Fragm. Phyt. Austr. i. 44.)

PALMÆ.

ARECA SAPIDA.

Solander in G. Forst. Plant. Esculent. Insul. Ocean. Austr. Comment. Botan.
66 ; G. Forst. Prodrom. 592 ; J. Hook. Flor. Nov. Zeel. i. 262, t. 59 ; A.
Banksii, Martius, Gen. et Spec. Palmar. 172.

Common on Pitt-Island, but rare on Chatham-Island.
The fruit responds to the description of the New Zealand palm.
The rachis of the leaves is tomentose.

Dr. Hooker justly observes, that A. sapida presents the most
southern member of the noble order of Palms. In East Gipps-Land
the species second in its extent to the south occurs, Livistona Australis; and more, there at a latitude of 37° 30' S. this majestic palm
raises itself still to a height of fully 80'.

JUNCEÆ.

LUZULA CAMPESTRIS.

Candolle et Lamark, Flore Française, iii. 161; R. Brown, Prodrom. 591 ;
Bicheno in Transact. Linn. Soc. xii. 334, t. 9, fig 4 ; E. Meyer, Synops.
Luzular. 17 ; Harpe, Essai d'une Monogr. des Joncées, 88 ; Spach, Suites
à Buffon, t. 113; Kunth, Enumer. iii. 307 ; E. Meyer in Lehm. Plant.
Preiss. ii. 48 ; J. Hook. Flor. Nov. Zeel. i. 264; Fl. Tasm. ii. 68 ; Steud.
Glumac. ii. 293 ; Benth. Handbook of the Brit. Flora, 541 ; L. picta, Ach.
Rich. Voy. de l'Astrolabe, i. 147 ; L. Banksiana, E. Meyer in Linnæa, xx.
412 ; L. crinita, J. Hook. Flor. Antarct. 85, t. 48; L. Oldfieldii, J. Hook.
Flor. Tasm. ii. 68 ; Juncus campestris, Linn. Spec. Plant. 468 ; Luciola
campestris, Smith, English Flora, ii. 181.

Chatham-Island on open places.
The species is widely diffused through temperate Australia, but
has not yet been found in our tropical or even subtropical regions.
All the plants of the genus contained in our extensive Australian
collection are referable to L. campestris, and to this species a host
of plants appearing under various specific names in the works on
the plants of different countries and resting on unreliable characters
are to be restored.

Sir James Smith in lucidly discussing the derivation of the word Luzula gives good reasons for suppressing it in favor of the more classical word Luciola.

A new circumscription of the limited number of genuine species of this genus, carried out with the aid of material such as the principal state-herbaria of Europe afford, appears highly recommendable.

JUNCUS PLANIFOLIUS.

R. Brown, Prodrom. Flor. Nov. Holl. 259; E. Meyer, Synops. Juncor. 36, and in Linnæa, iii. 370. & xxvi. 244; Harpe in Mémoir. de la Soc. d'Hist. Paris, ii. 55; Kunth, Enum. Plant. iii. 344; J. Hook. Flor. Autarctic. 358 & 545; Fl. Nov. Zeel. i. 263; Fl. Tasm. ii. 64; Steud. Synops. Plant. Glum. ii. 304.

Pitt-Island.

CYPEROIDEÆ.

HELEOCHARIS PALUSTRIS.

R. Brown, Prodrom. p. 80; H. acuta, R. Br. l. c.; H. gracilis, R. Br. l. c.; J. Hook. Flor. Nov. Zeel. i. 270; Flor. Tasman. ii. 85; H. multicaulis, Smith, English Flora, i. 64; H. uniglumis, Schultes, Mantissa Syst. Veget. ii. 83; H. mucronulata, Nees in Jard. Annal. and Magaz. of Natur. History, vi. 46; Scirpus palustris, Linn. Spec. Plant. 70; S. multicaulis, Smith, Flor. Britan. 48; S. uniglumis, Link, Jahrbuech. der Gewaechskunde, iii. 77.

Swampy places of Chatham-Island.

A plant as variable as might be anticipated from its wide distribution, its common occurrence, and the variety of ground it occupies, imitating in the southern hemisphere habitually all the forms of the northern. The stem varies from a few inches to three feet high, also much in thickness and the strength of its striæ. The root is sometimes long-creeping, sometimes scarcely emitting any suckers. The spike is also very variable in length, form and color; the scales are blunt or acute; the hypogynous setæ vary from 4–8 in number and may be occasionally more elongated or much reduced in length. The caryopsis alters in form, color and size, and so the persistent portion of the style.

The circumstance, that this plant in the northern hemisphere very prevailingly produces a bifid style, whilst in the southern hemisphere it occurs very predominantly with a trifid style, has been the cause of assigning to our plant specific rank. But in Arnhem's Land and on the Wimmera the author observed the bifid or trifid style promiscuously occurring in flowers of the same spike; and as

moreover from the material examined on this occasion it does not
appear that H. multicaulis can be maintained as specifically distinct,
we may consider that plant the tristigmatous representative of H.
palustris in the northern countries. The caryopsis of H. multicaulis
is often strongly triangular, as would generally be the case from the
development of three stigmata; but this character, on which this
supposed species mainly rests, appears to be one of degree. Indeed
its validity has been questioned by many meritorious observers. In
the Dukedom of Schleswig the author had repeated opportunities
of proving the characters derived from the bifid or trifid style of
Scirpus glaucus and S. lacustris to be fallible, and to trace the tran-
sitions between these two rushes. Heleocharis palustris, notwithstanding its producing in E. Australia almost invariably a trifid style,
shows singularly enough usually the normal pyriform- or obovate-
lenticular caryopsis of H. palustris, but not the triangular form of
that of H. multicaulis, but the two edges of the fruit may be more
or less obtuse or acute. To collect the scattered synonymy and the
quotations of the literature referring to this almost cosmopolitan
plant, would occupy pages and is beyond the scope of this publication.
H. acuta, though not admitted into Dr. Hooker's Tasmanian Flora,
will not likely prove different from our plant, as no allied plant has
been refound in Tasmania, to which the very brief phrase furnished
in R. Brown's Prodromus could be applied.

DEMOSCHŒNUS SPIRALIS.

J. Hook. Flor. Nov. Zeel. i. 272; Isolepis spiralis, Ach. Richard, Voy. de
l'Astrolabe, 105, t. 19; Anthophyllum Urvillei, Steudel, Synops. Plantar.
Glumacear. ii. 160.

Sandy places on the sea-shore of Chatham-Island.

The fruit-clasping scales are almost orbicular with truncate base.
The flat linear filaments are placed closely together on the convex
side of the caryopsis; the latter is exactly plane-convex.

Dr. Steudel by some oversight attributes to this plant a bifid
style and one-flowered spikelets.

CAREX PANICULATA.

Linn. Spec. Plant. 1383; Benth. Handbook of the Brit. Flor. 559; C. para-
doxa, Willd. Act. Academ. Berolin. 1794, 39, t. 1, fig. 1; C. teretiuscula,
Goodenough in Transact. Linn. Soc. ii. 163, t. 19, fig. 3; Boott in Jos.
Hook. Flor. Nov. Zeel. i. 221; C. appressa, R. Brown, Prodr. 242; Kunze,
Supplem. zu Schkuhr's Riedgräsern, tab. 11; J. Hook. Flor. Antarctic. i.

H

91; Nees in Lehm. Plant. Preiss. ii. 94; Raoul, Choix de Plant. de la Nouv. Zél. 40; Boott in J. Hook. Flor. Tasman. ii. 99; C. collata, Boott accord. to quotation in J. Hook. Fl. Nov. Zeel. i. 282; C. virgata, Banks & Soland. accord. to Boott in J. Hook. Flor. Nov. Zeel. i. 282; Illustr. of the Genus Carex, 46, t. 121 & 122; C. secta, Boott in J. Hook. Fl. Nov. Zeel. i. 281; Illustr. of the Gen. Carex, i. 47, t. 123 & 124; Vignea paniculata, V. paradoxa, V. teretiuscula, Reichenbach, Flor. German. Excursor, 409–410.

Open damp places of Chatham-Island.

The author has failed to discover any reliable characteristics by which the small forms of the Australian C. appressa could be safely distinguished from certain states of C. paniculata. Indeed Dr. Boott identified one of the varieties occurring in New Zealand with C. teretiuscula, which again, as Mr. Bentham justly remarks, is not specifically distinguishable from C. paniculata. Dr. Hooker whilst publishing his Antarctic Flora acknowledged C. appressa as a native of New Zealand, where it assumes under peculiar climatical and perhaps geological conditions some singularly aberrant forms. There also and in Australia, where C. paniculata is not subjected to the frigor of a lengthened winter, it luxuriates in uninterrupted growth, and this must be assigned as the main cause of the vigor of the vegetation shown by this species in our mild southern latitudes, when not rarely it is producing panicles of fully two feet length and leaves and stems proportionally long. Such luxurious states of the plant are those brought from Chatham-Island by Mr. Travers. But in Australia the species is occasionally met with quite as small as in Europe.

CAREX FORSTERI.

Wahlenberg in Act. Acad. Holmens. 1803, 154; Willdenow, Spec. Plant. iv. 249; Schkuhr, Botan. Handbuch, Caric. ii. 44; Boott in J. Hook. Flor. Nov. Zeel. ii. 285; Steudel, Synops. Glumac. ii. 206; Boott, Illustr. of the Genus Carex, i. 52, t. 137; C. debilis, Forst. Prodr. 550; C. recurva, Schkuhr, Handbuch, Caric. 120, t. Z. N. N. fig. 84; C. punctulata, A. Richard, Voy. de l'Astrolabe, 119, t. 22.

On damp open or bushy places of Chatham-Island.

The leaves are attaining $\frac{1}{4}''$ width. The number of spikes is sometimes increased to 10, of which then 5 are absolutely staminiferous; the female spikes are occasionally lengthened to 4″. Arista of bracteoles not rarely of considerable length. Fruit attaining a breadth of 1‴ and then verging into a roundish form.

Forster's name for this plant would receive precedence were it not singularly inapplicable to this robust species.

GRAMINE.Æ.

FESTUCA LITORALIS.

Labill. Nov. Holl. Plant. Specim. i. 22, t. 27 ; R. Brown, Prodrom. i. 178 ;
Poiret, Encyclop. Suppl. ii. 639 ; Ach. Rich. Voy. de l'Astrol. i. 123 ;
Kunth, Enum. Plant. i. 409 ; ii. 340 ; D. Dietr. Synops. Plant. i. 379 ;
J. Hook. Fl. Tasm. i. 128 ; F. scoparia, J. Hook. Flor. Antarctic. i. 99 ;
Fl. Nov. Zeel. i. 308 ; Schedonorus litoralis, Beauvois, Essai d'une Nouvelle
Agrostographie, 99 ; Rœm. et Schult. Syst. Veget. ii. 707 ; J. Hook. Flor.
Nov. Zeel. i. 310 ; Triodia Billardierii, Sprengel, Syst. Veget. i. 330 ;
Arundo triodioides, Trinius, Spec. Gramin. iii. t. 351 ; Steudel, Synops.
Glumac. i. 194.

On moist places of Chatham-Island.

In Australia this plant is restricted to the sandy coast, where it
is very abundant along a great extent of the extratropical shores
northward at least as far as Moreton Bay. By its creeping root it
aids in the retention of the sand.

The plant of Mr. Travers's collection has the glumellæ long-
bearded towards the base and toothless at the apex ; nor is the
character of the tridenticulated glumella always apparent in the
Australian plant.

This species mediates, as already indicated by R. Brown, the
transit to Triodia, and at least some species of this genus are in-
separable from Festuca ; for instance, Festuca irritans (Triodia irritans, R. Br. Prodr. 182), the Porcupine-Grass of the colonists and
the Spinifex of many of the Australian explorers, a plant widely
dispersed through the extratropical sandy desert, reaching to near
the shores at St. Vincent's and Spencer's Gulf ; this species has the
lower glumella terminated by three but exceedingly short teeth, of
which the middle one is almost obliterated.

Festuca viscida (Triodia pungens, R. Br. Prodr. 182 non Beauv. ;
T. viscida, Rœm. et Schult. Syst. Vegetab. ii. 599) is of common
occurrence on the sandstone tablelands of North Australia, and extends to the islands of the Gulf of Carpentaria. This species has
the lower glumella much deeper toothed than that of the preceding
congener, to small forms of which it is similar in external appearance ;
the teeth being from one-third to one-sixth of the length of the
glumella and acute. T. procera (R. Br. Prodr. 182) seems a variety
of F. viscida according to the diagnosis, to which some specimina
from the rocky desert-hills at the sources of the Victoria-River sufficiently respond. The length of the glumella and the degree of its
pubescence are variable.

F. microstachya (Triodia microstachya, R. Br. Prod. 182) may
notwithstanding its peculiar inflorescence and the extreme shortness
of the teeth of the glumellæ prove merely a variety of F. viscida.
It also occurs towards the sources of the Victoria-River.

Triodia Kerguelensis (J. Hook. Flor. Antarctic. ii. 379, tab. 138)
and T. antarctica (J. Hook. l. c. 380) are like the above-mentioned
Triodiæ referable to Festuca, and for these the appellations F. Kerguelensis and F. antipoda may be adopted.

AGROSTIS SOLANDRI.

A. Billardierii et A. æmula, R. Br. Prodr. 171 & 172; J. Hook. Flor. Tasm. ii.
115; A. avenacea, Gmel. Syst. Natur. i. 171; A. debilis, Poir. Encycl. i.
249; A. retrofracta, Willd. Enum. Plant. Hort. Berol. 94; A. Forsteri,
Rœm. & Schult. Syst. Veget. ii. 359; A. Rich. Voy. de l'Astrolabe, i. 131;
A. pilosa, A. Rich. l. c. 132, t. 23; Avena filiformis, Forst. Prodrom. 46;
Labill. Nov. Holl. Plant. Specim. i. 24, t. 31; Deyeuxia Billardierii, Kunth,
Révision des Gramin. i. 77; Schlechtend. Linnæa, xx. 564; J. Hook. Flor.
Nov. Zeel. i. 298; D. Forsteri, Kunth, Révis. i. 77; J. Hook. Flor. Nov.
Zeel. i. 298; D. æmula, Kunth, Révis. i. 77; Schlechten. Linnæa, xx. 565;
D. retrofracta, Kunth, Révis. i. 77; Lachnagrostis Forsteri et L. Billar-
dierii, Trinius de Gramin. Uniflor. 217; L. filiformis, Trin. Fundam. Agro-
stograph. 128; L. Willdenowii, Trin. de Gram. Uniflor. 217; L. retro-
fracta, Trin. Fund. 128; L. Preissii, Nees in Lehm. Plant. Preiss. i. 97;
Calamagrostis æmula et C. Willdenowii, Steudel, Glumac. i. 192.

On moist ground of Chatham-Island.

In Australia widely dispersed through the extratropical tracts.

The adoption of a new specific name for this grass needs hardly
an explanation. R. Brown expressively excluded from A. Billardierii
Forster's plant, and like all subsequent authors did neither this
illustrious observer recognize in Australia its true limits, which even
now in the given synonymy are but imperfectly circumscribed.
Gmelin's and Poiret's names are apt to impress a wrong notion of
the plant. Under such circumstances it was deemed advisable to
bestow on this grass the name of the distinguished companion of Sir
Jos. Banks, who not only was the first discoverer of this species, but
who also at once recognized its true generic position, although he
again adopted for it more than one specific appellation.

The habit and leaves of this grass are almost as vacillate as
those of Poa Australis or Danthonia penicillata. The penicillar
rudiment of a second flower is very variable in length and often
especially in small-flowered individuals reduced to a minute beard.

Holcus lanatus.

Linné, Spec. Plant. 1485 ; Smith, Engl. Flora, i. 108.

Open places of Chatham-Island.

Evidently introduced there like into many parts of Australia.

Arundo conspicua.

Forster, Prodrom. 48, accord. to J. Hook. Flor. Nov. Zeel. i. 299 ; A. australis, Rich. Voy. de l'Astrolabe, i. 121; A. Cunn. in Hook. Compan. to the Bot. Mag. ii. 371 ; A. Kakao, Steud. Synops. Glumac. i. 194.

Open damp places of Chatham-Island, where it is called "Toi Toi" by the natives.

The spikelets produce sometimes a fourth flower, which then is imperfect. The subulate teeth at the apex of the glumella are variable in length.

It needs further investigation for ascertaining whether any of the following plants are referable to A. conspicua : A. australis, Cavan. in Annal. de Cienc. accord. to Rœm. et Schult. Syst. Veg. ii. 511 ; A. Egmontiana, Rœm. & Schult. l. c. ; Calamagrostis conspicua, Gmelin, Syst. Nat. i. 172 ; Agrostis conspicua, Rœm. et Schult. Syst. Veg. ii. 364 ; Achnatherum conspicuum, Beauv. Essai, 20 ; Gynerium Zeelandicum, Steud. Glumac. i. 198.

The restriction of this Reed to New Zealand and the small adjacent islands is a curious fact in phytogeography, when contrasted with the wide distribution of A. Phragmites over the globe. The latter plant is also a native of New Zealand, according to a specimen brought by Dr. Haast from the Grey-River.

Aira cæspitosa.

Linné, Spec. Plant. 96; Smith, English Flora, i. 103 ; Koch, Synops. Flor. German. et Helvet. ii. 914; Steud. Glumac. i. 218; A. Kingii, J. Hook. Flor. Antarctic. 376, t. 135 ; Deschampsia cæspitosa, Beauv. Essai d'une Nouv. Agrostographie, 91, t. 18, f. 3 ; Kunth, Enum. i. 286 et Suppl. 241 ; J. Hook. Flor. Nov. Zeel. i. 301 ; Fl. Tasm. ii. 118.

On moist ground of Chatham-Island.

Found also in various parts of New Zealand, for instance in the Province of Canterbury, in Tasmania, and in Australia from the Tattiara-country to Gipps-Land, ascending to the higher parts of the Australian Alps.

The spikelets of the Chatham-plant are rather larger than those of the ordinary European state of this grass ; but somewhat smaller than those of the antarctic South American plant. Their pedicellar rudiment of a third flower is strongly developed, whilst in some specimina collected in Australia and New Zealand it bears glumellæ, rendering thus the spikelet truly triflorous.

LYCOPODINEÆ.

LYCOPODIUM VOLUBILE.

Forst. Prodrom. 482 ; Swartz, Synops. Filic. 180 & 404; Hook. et Greville, Icon. Filic t. 170; Spring, Mémoir. Acad. Bruxell. xv. 105 ; J. Hook, Flor. Nov. Zeel. ii. 55.

Common on Chatham-Island.

LYCOPODIUM DENSUM.

Labillard. Nov. Holl. Plant. Specim. ii. 104, t. 251, fig. 1 ; Willd. Spec. Plant. v. 22 ; R. Brown, Prodrom. 165 ; Spring, Mémoir. Acad. Bruxell. xv. 87 ; J. Hook. Flor. Nov. Zeel. ii. 53 ; Flor. Tasm. ii. 156.

Found in Chatham-Island by Dr. Dieffenbach.

LYCOPODIUM SELAGO.

Linné, Spec. Plant. 1565 ; J. Hook. Flor. Antarct. ii. 394 ; Flor. Nov. Zeel. ii. 52 ; Flor. Tasman. ii. 155, tab. 170 A.; L. varium, R. Br. Prodr. 164; Hook. & Grevill. Icon. Filic. 112 ; J. Hook. Flor. Antarctica, i. 115 ; Flor. Nov. Zeel. i. 52 ; Flor. Tasm. i. 155, tab. 170, B.-F.; L. Flagellaria, Rich. Voy. de l'Astrolabe, i. 60; L. Billardierii, Spring in Mémoir. Acad. Brux. xv. 56 ; J. Hook. Flor. Nov. Zeel. ii. 53.

Varietas flagellaria.

Pitt-Island, rare on rocks.

Our collection contains from Stuart-Island specimina exhibiting strictly bracteate and foliate spikes on the same individual plant.

In Australia the typical L. Selago is restricted to the highlands ; the elongated varieties are observed in deep moist forest-ravines either on fern-trees or on mossy rocks.

FILICES.

GLEICHENIA DICARPA.

R. Br. Prodrom. 161; Kunze, Farnkræuter, 164, t. 70, fig. 2 ; Hook. Spec. Filic. i. 3, t. 1, C. ; Exotic Ferns, 40 ; J. Hook. Flor. Nov. Zeel. ii. 5 ; Flor. Tasm. ii. 131 ; G. alpina, R. Br. Prodr. 161 ; Hook. et Grev. Icon. Fil. t. 58.

Chatham-Island.

Dr. Hooker (Flor. Nov. Zeel. ii. 5) records G. microphylla from Chatham-Island. The normal form of that fern occurs not in Mr. Travers's collection; but the variety hecistophylla from New Zealand seems to obliterate the limits of these two species. If so we possess in Australia but four Gleicheniæ, viz. G. Platyzoma (Platyzoma microphyllum, R. Br. Prodr. 160), which is widely dispersed over the arid interior of tropical Australia; G. microphylla, of which G. rupestris and G. speluncæ are varieties; G. flabellata and G. Hermanni. The reduction of Platyzoma to Gleichenia has been suggested by Prof. Endlicher (Gen. Plant. 64) and Sir Will. Hooker. The development of barren short and narrow fronds accessory to the fertile ones is its principal character.

BOTRYCHIUM TERNATUM.

Swartz, Synops. Filic. 172; Willd. Spec. Plant. v. 63; Kunze, Farnkræuter, t. 121; B. lunaroides, Swartz, l. c.; Torrey, Flora of New York, ii. 506; Chapman, Flora of South Unit. States, 599; B. dissectum, Spreng. Anleitung, iii. 172; B. obliquum, Muehlenb. in Willd. Spec. Plant. v. 63; B. fumaroides, Willd. l. c.; B. matricarioides, Willd. l. c.; B. rutaceum, Svensk Bot. t. 372, fig. 2; B. Fumariæ, Spreng. Syst. Veg. iv. 23; B. Matricariæ, Spreng. l. c.; B. Australe, R. Br. Prodrom. 164; B. rutifolium, Al. Br. in Dœlling. Rhein. Flor. 24; B. Virginicum, J. Hook. Flor. Nov. Zeel. ii. 50; Flor. Tasman. ii. 154, t. 169; Osmunda Lunaria var. Bæckeana, Linn. Amœn. Acad.; O. ternata, Thunb. Flor. Japon. 329, t. 32; O. Lunaria, Oeder in Flor. Danic. t. 18, fig. 2; O. Matricariæ, Schrank, Flor. Bavar. ii. 419; O. biternata, Lam. Encycl. iv. 608; Botrypus lunaroides, Michaux, Flor. Amer. Boreal. ii. 274.

On open shady places of Chatham-Island.

This fern is very rare in South Australia and in the lowlands of Victoria and New South Wales, and not yet found in tropical and Western Australia, but not unfrequent in some of the highlands of this continent. Some specimina gathered by Dr. Haast in the Province of Canterbury, New Zealand, show two sterile fronds.

It remains to be ascertained in what relation B. ternatum is standing to B. Lunaria and B. Virginicum. Professor Oeder, an excellent observer, illustrated (Flora Danica, t. 18) both B. Lunaria and B. ternatum as varieties of one species, connected by B. rutaceum. Smith (Engl. Flora, iv. 315) and Weber & Mohr (Bot. Taschenbuch, 48–51) adopt this view, in which the author after observing the variability of these plants in the alpine valley of the Cabongra is inclined to concur. Wallroth (Compend. Flor. Germ. iii. 30) and

Lange (Dansk Flora, 687) include B. rutaceum in B. Lunaria, but uphold B. Matricariæ as a species, whilst Alex. Braun (in Dœlling. Rhein. Flor. 24) and Koch (Synops. Flor. German. iii. 972) regard all three plants as specifically different.

B. Virginicum quoted by Dr. Hooker as identical with B. ternatum is kept distinct by Torrey and Chapman, and has as yet not been found in any part of Australia, but occurs in Russia, specimina collected near Petersburg and communicated by Dr. Regel fully corresponding to American ones gathered in Connecticut and sent by Prof. A. Gray. For the illustration of this plant Plukenet's figure (Amalth. t. 427, fig. 8) is evidently intended. B. daucifolium (Wall. in Hook. & Grev. Icon. Filic. t. 161 ; Bot. Mag. 5, 340) and B. subcarnosum, both received from Mr. Thwaites as gathered in Ceylon, may mediate a transit from B. Virginicum to B. ternatum. Dr. Hooker is inclined to refer all known Botrychia to two species. To the writer of these pages it seems by no means improbable, that all are the offsprings of one species, in which three primary varieties may be recognized.

TRICHOMANES VENOSUM.

R. Brown, Prodr. 159 ; Hook. & Grev. Icon. Filic. t. 78 ; Hook. Spec. Filic. i. 132 ; J. Hook. Flor. Nov. Zeel. ii. 17 ; Flor. Tasm. ii. 135.

On mossy stems of fern-trees of Chatham-Island.

In the Australian Continent limited to the south-eastern ferntree country.

HYMENOPHYLLUM DEMISSUM.

Swartz, Synops. Filic. 147 et 374 ; Schkuhr, Farnkræuter, 135, t. 135, C. ; Willd. Spec. Plant. v. 529 ; A. Rich. Voy. de l'Astrolabe, i. 92 ; Hook. Spec. Filic. i. 109 ; J. Hook. Flor. Nov. Zeel. i. 14 ; H. flabellatum, Labill. Nov. Holl. Plant. Specim. ii. 101, t. 250, B. ; R. Br. Prodr. 159 ; J. Hook. Fl. Nov. Zeel. ii. 15 ; Fl. Tasm. ii. 134 ; H. nitens, R. Brown, Prodr. 159 ; Hook. et Grev. Icon. Filic. t. 197 ; Trichomanes demissa, Forst. Prodr. 468.

On mossy stems of fern-trees of Chatham-Island.

In Australia restricted to the south-eastern fern-tree gullies.

Our collections are not sufficiently extensive to offer from their investigation a complete synonymy of this species, but they are sufficiently convincing of the fact that the islands of New Zealand possess very few and Australia probably but two real species of this genus. The winged rachis and stipes afford no positive mark for specific discrimination and perhaps neither the indument. The specific names given by Swartz and Labillardière simultaneously to this lovely plant are not well applying to the generality of its forms.

CYATHEA DEALBATA.

Swartz, Synops. Filic. 140 & 365 ; Willd. Spec. Plant. v. 495 ; Ach. Rich. Voy. de l'Astrolabe, 77, t. 10 ; Hook. Spec. Filic. i. 27 ; J. Hook. Flor. Nov. Zeel. ii. 7 ; Ralph in Proceed. Linn. Soc. iii. 163 & 164 ; Polypodium dealbatum, Forst. Prodr. 454.

Common in woods of the Chatham-Islands.

The serratures or crenatures of the ultimate lobes of the frond are often very faint.

CYATHEA CUNNINGHAMI.

J. Hook. in Hook. Icon. Plant. 985 ; Flor. Nov. Zeel. ii. 7 ; Ralph in Proceed. Linn. Soc. iii. 163 & 167.

Common in the woods of the Chatham-Islands, where it is called by the natives "Ponja."

If to the plants in our collection a right interpretation has been given this species differs from C. medullaris besides in some of the characters pointed out by Dr. Hooker in slightly smaller more tender sori, which seem not so generally distributed over the whole but only along the lower part of the underpage of the pinnule-segments, in more slender sporangia with a less conspicuous ring.

C. Smithii, according to specimina from Canterbury, has generally but few sori attached to the midnerve of the lower part of each of the pinnule-segments, and the small indusium retracted to a shallow cup, from which the prominent receptacle comes conspicuously in view.

DICKSONIA SQUARROSA.

Swartz, Synops. Filic. 136 & 355 ; Schkuhr's Farrnkræuter, 124, t. 130 ; Hook. Spec. Filic. i. 68 ; J. Hook. Flor. Nov. Zeel. ii. 9 ; Ralph in Proceed. Linn. Society, iii. 169 ; Trichomanes squarrosum, Forst. Prodr. 476.

Chatham-Islands.

The specimina of this Dicksonia before us are scarcely sufficient for exact identification.

ASPLENIUM FALCATUM.

Lamark, Encycl. Méthod. ii. 303 (1786); Swartz, Flor. Ind. Occid. iii. 1618 ; Syn. Fil. 77 ; Retzius, Observ. Fasc. 6, 37 ; Willd. Spec. Plant. v. 325 ; R. Br. Prodr. Flor. Nov. Holl. 150; A. Rich. Voy. de l'Astrolabe, i. 73 ; Endl. Prodrom. Flor. Ins. Norfolk. 9 ; Hook. Spec. Filic. iii. 160 ; A.

I

polyodon, G. Forst. Prodr. 428 (1786); J. Hook. Flor. Nov. Zeel. ii. 34;
A. cultratrum, Gaudich. in Freycen. Voy. Bot. 317; A. Forsterianum,
Colenso in Tasman. Phil. Journ. accord. to Jos. Hook. Flor. Nov. Zeel. ii.
34; Trichomanes adiantoides, Linn. Spec. Plant. 1561; Flor. Zeylanic.
385; Burmann, Flor. Indic. 236; Tarachia polyodon, Presl, Epimel. 76.

Chatham-Island, in woods on trees and in rocky crevices.

ASPLENIUM MARINUM.

Linné, Spec. Plant. 1540.

Varietas obtusata.

A. obtusatum, Forst. Prodrom. 430; Schkuhr, Farrnkr. t. 68; Labill. Nov.
Holl. Plant. Specim. ii. 93, t. 242, f. 2; J. Hook. Fl. Nov. Zeel. ii. 33;
Hombr. & Jacquem. Voy. au Pôle Sud, Bot. Crypt. t. 1; Hook. Exotic Ferns,
t. xlvi.; J. Hook. Flor. Tasm. ii. 145; Hook. Spec. Filic. iii. 96; A. obliquum, Forst. Prodr. 429; Labill. l. c. t. 242, fig. 1; Schkuhr, Farrnkr. t.
71; A. difforme, R. Br. Prodr. 160; Endl. Prodrom. Flor. Insul. Norfolk.
9; A. oblongifolium, Colenso in Tasm. Phil. Journ. sec. J. Hook. l. c.; A.
apicidentatum, Hombr. & Jacquem. l. c.

On rocky places of Chatham-Island almost everywhere. Not
rare in New Zealand and Tasmania, but very scarce in Continental
Australia. This variety is distinguished by more or less coriaceous,
simply pinnate not very elongated fronds, by often blunt finely or
grossly serrate-crenate or towards the base pinnatifid or pinnatisected
pinnæ and by rather long sori.

Our collections contain of this variety specimina only 2–3" high
with ample fructification; further forms, which plainly connect this
plant with A. lucidum and with A. flaccidum.

Varietas bulbifera.

A. bulbiferum, Forst. Prodr. 433; Schkuhr, Farrnkr. 74, t. 79; Hook. Icon.
t. 423; Hombr. & Jacquem. l. c. t. 3, fig. 1; J. Hook. Flor. Nov. Zeel. ii.
34; Fl. Tasm. ii. 146; Hook. Spec. Fil. iii. 196; A. laxum, R. Br. Prodr.
151; A. viridans, Labill. Sert. Austro-Caled. i. 2, t. ii.; A. triste, Raoul,
Choix, p. 10.

Common in woods of the Chatham-Islands on the border of
water-courses. Frequent in New Zealand, Tasmania and in the fern-
tree gullies of South-East Australia.

This variety is characterized by large more or less membranous
bipinnate fronds, acuminate pinnæ, pinnatisected or pinnatifid lower
pinnules and short sori.

Some of the viviparous specimina of our collections produce from
near the apex of the pinnæ young plants fully 2–3" long.

Varietas flaccida.

A. flaccidum, Forst. Prodr. 426 ; J. Hook. Fl. Antarct. i. 109 ; Fl. Nov. Zeel.
ii. 35 ; Fl. Tasm. ii. 146 ; Hook. Spec. Filic. iii. 205 ; A. odontites, R. Br.
Prodr. 151 ; A. heterophyllum, A. Rich. Voy. de l'Astrolabe, i. 74 ; A. ap-
pendiculatum, Presl, Tentam. Pteridogr. ; Sond. & Muell. in Linnæa, xxv.
718 ; Cænopteris flaccida, Thunb. Nov. Act. Petropol. ix. 158, t. D, fig. 1
& 2 ; Schk. Farrnkr. t. 82 ; Spreng. Syst. Veg. iv. 90 ; C. odontites, Thunb.
l. c. t. E, fig. 1 ; Spreng. Anleit. iii. 115, t. 3, fig. 24 ; C. appendiculata,
Labill. l. c. ii. 94, t. 243 ; Darea flaccida, Smith, Memoir Acad. Turin. v.
409 ; Darea odontites, Willd. Spec. Plant. v. 296 ; D. flaccida, Willd. l. c.
295.

Common on trees of the Chatham-Islands.

Abundant in New Zealand and Tasmania and on fern-trees of
South-East Australia.

This variety is known by its more or less coriaceous or even
carnulent simply or doubly pinnated fronds, by pinnatifid or pinnatisected acuminate pinnules with rather distant lobes, by generally
rather short occasionally elongated mostly marginal sori.

Middle forms between this variety and A. bulbiferum and A.
Richardi exist in our collections. The latter, occurring in clefts of
mountain-rocks of New Zealand, must be regarded also as a variety,
remarkable for the often small size of its fronds and for its pinnules
being cut into fine narrow or short lobes, bearing marginal sori.
Neither this variety nor those represented by A. lucidum and A.
adiantoides occur in Mr. Travers's Chathamian collection, nor are
they known to exist in Australia or Tasmania. A. adiantoides, according to notes by Dr. Haast, grows from the fissures of mountain-
rocks, and thus its locality may account for the remarkable aberration
of the form and size of its pinnules. Middle forms between it and
A. obtusatum are preserved in our collection. Clear transits from
A. lucidum to A. obtusatum have also been observed, whilst a form
with pinnatifid pinnæ, brought by Dr. Haast from the limestone-
rocks of the coast between the Buller- and Grey-River, approaches
to certain less divided states of A. bulbiferum.

If in reality specific differences exist between any of the plants
here referred to A. marinum, such must emanate in characters
different from any of those hitherto pointed out. The reason of the
greater variability of this fern in New Zealand than even in Australia must be sought in climatical and geological conditions.

Some specimina of A. marinum, collected by Sieber in Corsica
and others gathered in Spain by Prof. Lange, approach so near to

certain states of A. obtusatum, that no specific value to their characters can possibly be attached. It also appears probable to the
author, that what he has drawn together on this occasion as forms
of this fern, does not even exhaust the display of its variations. In
the ferntree-ravines of Victoria the writer had frequent occasion to
trace A. bulbiferum and A. flacciduin into each other.

POLYPODIUM GRAMMITIDIS.

R. Br. Prodr. 147; J. Hook. Flor. Antarctic. i. 111; Fl. Nov. Zeel. ii. 41;
    Fl. Tasm. ii. 150; Hook. Spec. Filic. iv. 230; Mettenius, Polypod. 53;
    Grammitis heterophylla, Labill. Nov. Holl. Plant. Specim. ii. 91, t. 239;
    Xiphopteris heterophylla, Spreng. Syst. Veg. iv. 44.

On trunks of trees in woods of the Chatham-Islands. Not rare
in New in Zealand, Tasmania and the ferntree-country of South-East
Australia.

In a variety, which might be called serriformis, occurring in New
Zealand, the fronds are not pinnatisected, but simply linear and
toothed in the manner of a segment of the ordinary frond. The
form of the sori of Polypodium Australe is not unfrequently alike
to those of Polypodium Grammitidis. Thus that plant links the
genera Grammitis and Polypodium closely together. It is a frequent
companion of P. grammitidis in Australia felix, and ascends in New
Zealand to the glacier-regions. The fronds of P. grammitidis are
sometimes fully 1' long.

POLYPODIUM PENNIGERUM.

Forst. Prodr. 444; Schkuhr, Farrnkræut. 17, t. 22; Hook. Spec. Filic. 7;
    Aspidium pennigerum, Swartz, Syn. Filic. 49 & 250; Ach. Rich. Voy. de
    l'Astrol. i. 67; Nephrodium pennigerum, Desvaux, according to Hook. l. c.;
    Goniopteris pennigera, J. Smith, Gen. Ferns, p. 18; J. Hook. Fl. Nov.
    Zeel. i. 40.

Common in Chatham-Island on borders of water-courses.

Pinnæ not seldom alternate.

POLYPODIUM RUGOSULUM.

Labill. Nov. Holl. Plant. Specim. ii. 92, t. 241; Willd. Spec. Plant. v. 1206;
    R. Brown, Prodr. 147; Endlich. Prodr. Fl. Ins. Norf. 7; J. Hook. Flor.
    Nov. Zeel. ii. 41; Flor. Tasm. ii. 149; Hook. Spec. Filic. iv. 272 (with
    extensive synonymy); Beddome, Ferns of South India, 56, t. 170; P. vis-
    cidum, Spreng. Syst. Veg. iv. 61; J. Hook. Flor. Antarctic. 110; Chei-
    lanthes ambigua, Ach. Rich. Voy. de l'Astrolabe, i. 84.

Common in woods of the Chatham-Islands.

Mr. Travers's plant represents the normal form.

Some varieties of this fern, as indicated by Dr. Hooker, bear great resemblance to the exindusiate form of Nephrodium velutinum, others to Hypolepis tenuifolia.

POLYPODIUM SCANDENS.

Forst. Prodr. 437; Schkuhr, Farrnkr. 11, t. 8; Swartz, Syn. Fil. 131 & 228; Willd. Spec. Plant. v. 166; P. pustulatum, Forst. Prodr. 436; Schk. Farrnkr. 11, t. 10; Bernhardi in Kœnig & Sims's Annal. t. 1, fig. 6; Metten. Polypod. p. 101; Hook. Spec. Filic. v. 80.; P. membranifolium, R. Br. Prodr. 147; P. phymatodes, Rich. Voy. de l'Astrol. 64; Phymatodes pustulata, Presl, Tentamen Pteridograph. 196; J. Hook. Flor. Nov. Zeel. ii. 42; Drynaria pustulata, J. Smith, accord. to Hook. l. c.

Common in woods of Chatham-Island. To be found also in New Zealand and widely through tropical and extratropical East Australia.

Varietas Billardierii.

Polypodium scandens, Labill. Nov. Holl. Plant. Specim. ii. 91, t. 240; P. diversifolium, Willd. Spec. Plant. v. 166; P. Billardierii, R. Br. Prodr. 147; Metten. Polypod. 101; Endl. Prodr. Flor. Insul. Norfolk. 7; Hook. Spec. Filic. v. 147; Phymatodes Billardierii, Presl, Tentam. Pteridogr. 196; J. Hook. Flor. Antaret. i. 111; Flor. Nov. Zeel. ii. 42; Flor. Tasm. ii. 150; Drynaria Billardierii, J. Smith.

Common everywhere in the Chatham-Islands, also in New Zealand, Tasmania, and in Australia from the southern part of Queensland through New South Wales and the colony of Victoria to the eastern part of South Australia.

The author is unable to concede specific value to the differences hitherto drawn between the two plants here recorded. It requires even still further investigations, whether reliable characters exist between P. scandens and P. phymatodes, plants perhaps rightly united by Achilles Richard. Several other described Polypodia seem referable to our plant.

The identification of Forster's P. scandens from the Pacific islands with P. pustulatum rests on the authority of Mettenius. The name of the Society-Island's plant has been chosen in preference to that simultaneously bestowed on the New Zealand plant, being exquisitely appropriate.

NEPHRODIUM DECOMPOSITUM.

R. Brown, Prodr. Flor. Nov. Holl. 149; J. Hook. Flor. Nov. Zeel. ii. 39, t. 79; Fl. Tasm. i. 149; Hook. Spec. Filic. iv. 146; N. microsorum, Endl. Prodr.

Flor. Insul. Norfolk. 9 ; N. glabellum, A. Cunningh. in Hook. Compan. to the Bot. Mag. ii. 367 ; Metten. Aspid. 69 ; Aspidium Shepherdi, Kunze in Linnæa, xxiii. 230 ; A. acuminatum, Lowe, Filic. 6, t. 11 ; Metten. Aspid. p. 71 ; Lastræa acuminata, Th. Moore ; L. davallioides, Brackenr. Filic. Wilk. Unit. Stat. Explor. Exped. 202 ; L. atro-virens, J. Smith, Catal. Cult. Ferns, 59.

Common in woods of Chatham-Island. Not rare in New Zealand ; occurring also in a few places of Tasmania ; restricted in Australia to Gipps-Land, New South Wales and the southern part of Queensland ; also found in Norfolk-Island and some of the Pacific groups.

Bynoe's plant was most probably gathered in New South Wales.

Nephrodium hispidum (Hook. Spec. Fil. iv. 150) mediates the transit of this genus to Aspidium. That fern may be sought for yet in the Chatham-group.

ASPIDIUM CORIACEUM.

Swartz, Synops. Filic. 57 ; Schkuhr, Farrnkr. t. 50 ; R. Br. Prodr. 148 ; Willd. Spec. Plant. v. 268 ; Metten. Filic. Hort. Lipsiens. 89 ; Endl. Prodr. Flor. Insul. Norfolk. 8 ; Hook. Spec. Filic. iv. 32 ; A. Cunninghami, Colenso in Tasm. Phil. Journ. accord. to J. Hook. l. c. ; Polypodium adiantiforme, Forst. Prodr. 449 ; P. coriaceum, Swartz, Prodr. Descript. Vegetab. Ind. Occid. 133 ; Sw. Flor. Ind. Occid. iii. 1688 ; Polystichum coriaceum, Schott, Gen. Fil. ; J. Hook. Flor. Nov. Zeel. ii. 37 ; Fl. Tasm. ii. 148.

On tree-ferns of Chatham-Island not common. Not unfrequent in New Zealand, Tasmania and ranging in Australia from the vicinity of Cape Otway eastward to New South Wales.

The indusia are generally early seceding, and seemingly not always developed.

ASPIDIUM ACULEATUM.

Swartz, Synops. Filic. 53 ; Schkuhr's Farrnkr. t. 39 ; Smith, English Flora, iv. 277 ; Koch, Synops. Flor. Germ. et Helv. iii. 976 ; Benth. Handb. of the Brit. Fl. 628 ; Hook. Spec. Filic. iv. 18-22 (with extensive quotations of literature) ; A. vestitum, Sw. Syn. Filic. 53 & 254 ; Schk. Farrnkr. t. 43 ; A. lobatum, Sw. l. c. 53 ; A. proliferum, R. Br. Prodr. 147 ; Sond. & Muell. in Linnæa, xxv. 718 ; A. coriaceum var. acutidentatum, A. Rich. Voy. de l'Astrol. i. 71 ; A. venustum, Hombr. & Jacquem. Voy. au Pôle Sud, t. 5 ; J. Hook. Flor. Antarctic. i. 106 ; A. pulcherrimum et A. Waitkarense, Colenso in Tasm. Phil. Journ. accord. to J. Hook. l. c. ; A. Richardi, Hook. Spec. Filic. iv. 23, t. 222 ; Polypodium aculeatum, Linné, Spec. Plant. 1552 ; P. vestitum, Forst. Prodr. 445 ; P. silvaticum, Colenso in J. Hook. Flor. Nov. Zeel. ii. 41, t. 81 ; Polystichum aculeatum, Roth. Flor. Germanic. iii. 79 ; Beddome, Ferns of South India, 41, t. 121 ; P. vestitum,

Presl, Tentam. Pteridolog. 83 ; J. Hook. Flor. Nov. Zeel. ii. 38 ; Flor.
Tasm. ii. 148 ; P. aristatum, J. Hook. Flor. Nov. Zeel. ii. 37.

Common in woods and on the borders of water-courses of Chatham-Island ; extending in Continental Australia from Mount Gambier
to New South Wales, abounding in the moister forest-regions,
ascending to the highest alpine elevations, and forming in some of
our highlands, for instance on the Baw Baw Range, with Lomaria
Capensis, the predominant part of the lower vegetation. (Conf. F. M.
Report for 1860, p. 14.)

When, especially in colder forest-glens or along shady rivulets,
this fern occurs in a luxuriant state with fronds 4–6' long, its circular
tufts then assume a truly grand appearance. The chaff-like scales
are pale-brown in the Chatham-plant and the pinnules bluntly
crenate. The scales of the stipites of this species attain in richly
developed plants fully the length of 1", and impart when shining
black and brown margined to this fern great additional beauty.
The indusium readily secedes and seems occasionally to be wanting.

Diminutive varieties have been detected on the Ashburton-glacier
and in others of the permanently snowy regions of New Zealand by
Dr. Haast.

LOMARIA DISCOLOR.

Willd. in Magaz. der Naturf. Freunde zu Berlin, 1809, 160 ; Spec. Plant. v.
293 ; J. Hook. Flor. Nov. Zeel. ii. 30 ; Flor. Tasm. ii. 143 ; Hook. Spec.
Filic. iii. 5 ; L. nuda, Willd. Spec. Plant. v. 289 ; L. falcata, Spreng. Syst.
Veg. iv. 62 ; L. lanceolata, J. Hook. Flor. Antarctic. i. 110 ; Osmunda
discolor, Forst. Prodr. 413 ; Onoclea discolor, Swartz, Syn. Fil. 111 ; O.
nuda, Labill. Nov. Holl. Plant. Specim. ii. 96, t. 246 ; Hemionitis discolor,
Schk. Farrnkr. 7, t. 6 ; Stegania nuda, R. Br. Prodr. 153 (1810) ; S. falcata,
R. Br. l. c. ; S. discolor, Ach. Rich. Voy. de l'Astrol. ii. 87 ; S. procera,
A. Rich. l. c. t. 13, the barren frond.

In woods of Chatham-Island common. In Continental Australia
from St. Vincent's Gulf extending to New South Wales.

This and L. Capensis are the only ferns hitherto in the Australian
Continent found westward of Mount Gambier.

Fronds arranged in circles; the fertile segments often broad,
stout and blunt.

The fronds are larger and firmer and less saturated green than
those of L. lanceolata, which corresponds in color and texture of its
foliage to L. fluviatilis ; the segments moreover are neither distinctly
crenulated, nor the inferior ones so singularly abbreviated.

The segments of the fronds are not rarely along their upper part
fruit-bearing and in the lower portion barren.

A strange variety of L. discolor (var. bipiunatisecta) with segments cleft again deeply into lobes has been discovered in the Dandenong-Ranges of Australia felix by Mr. D. Boyle.

The absence of the allied genus Blechnum in New Zealand and its dependencies is remarkable.

LOMARIA CAPENSIS.

Willd. Spec. Plant. v. 291; Linnæa, 1847, 256; Rawson & Pappe, Enum. Filic. Cap. 27; L. striata, W. l. c.; L. lineata, W. l. c.; L. Chilensis, Kaulf. Enum. Filic. 154; Hook. Gen. Filic. t. lxiv. 13; L. procera, Spreng. Syst. Veg. iv. 65; Hook. Icon. 427; J. Hook. Flor. Antarctic. i. 110; Flor. Nov. Zeel. i. 27, t. 75; Flor. Tasmanic. i. 142; Hombr. & Jacquem. Voy. au Pôle Sud, t. 2, E; Hook. Spec. Filic. iii. 22–28 (with extensive synonyma); L. minor, Spreng. Syst. Veg. v. 291; L. latifolia, Colenso in Tasm. Phil. Journ. accord. to J. Hook. l. c.; Osmunda Capensis, Linné, Mantissa Plant. 306; O. procera, Forst. Prodr. 414; Onoclea Capensis, Swartz, Syn. Filic. 111; Blechnum procerum, Sw. Syn. Fil. 115; Labill. Nov. Holl. Plant. Specim. ii. 97, t. 247; Willd. Spec. Plant. v. 415; Asplenium procerum, Bernh. Act. Erfurt. 1802, 4, fig. 1; Stegania procera, R. Br. Prodr. 153; Achill. Rich. Voy. de l'Astrol. t. 13, the fertile frond; S. minor, R. Br. l. c.; Parablechnum procerum, Presl, Epimel. 109.

Common in forest-ground of Chatham-Island. In Continental Australia not rare from St. Vincent's Gulf to New South Wales, in some of the interjacent localities very abundant (conf. Linnæa, 1847, 561), but as yet not found in South-West Australia, where seemingly the genus Lomaria is not represented.

Decidedly the tallest of all Australian congeners and less exclusively a forest-plant than most other species, ascending to alpine elevations. Hence this noble fern might easily be naturalized in Middle Europe.

ADIANTUM FORMOSUM.

R. Br. Prodr. Flor. Nov. Holl. 155; Ach. Rich. Voy. de l'Astrolabe, i. 88; A. Cunn. in Hook. Compan. to the Bot. Mag. ii. 366; Raoul, Choix de Plant. de la Nouv. Zél. 38; Hook. Spec. Filic. ii. 51, t. 86, B; J. Hook. Fl. Nov. Zeel. ii. 21.

Varietas Cunninghami.

A. Cunninghami, Hook. Spec. Filic. ii. 52, t. 86, A; J. Hook. Flor. Nov. Zeel. ii. 21.

Rare in Chatham-Island on rocky damp places.

Specimina with slightly downy raches gathered in New Zealand obliterate the demarcation between A. Cunninghami and A. formosum, none also of the other characters relied on for discrimination proving constaht.

The normal form of A. formosum exists in Continental Australia from East Gipps-Land to the southern part of Queensland.

The closely allied A. affine, discovered also by Dr. Ludwig Leichhardt on Archer's Creek of subtropical Eastern Australia, has a root very different to that of A. formosum.

New Zealand specimina of this plant, distributed from Dr. Sinclair's collection by Mr. W. Gourlie, have the minute dark bristles much more copiously scattered over the lower pages of the pinnules than plants of either Australia or Norfolk-Island.

A. hispidulum ranges in Australia southward to the Genoa River.

PTERIS AQUILINA.

Linné, Spec. Plant. 1533; J. Hook. Flor. Nov. Zeel. ii. 25; Flor. Tasm. ii. 139; Hook. Spec. Filic. ii. 196; P. esculenta, Forst. Prodr. 79; Plant. Esculent. Insul. Ocean. Austr. 74; Labill. Nov. Holl. Plant. Specim. ii. 95, t. 244; R. Br. Prodr. 154; Blume, Enum. Filic. Jav. 214; Ach. Rich. Voy. de l'Astrolabe, ii. 79; Endl. Prodr. Flor. Insul. Norfolk. 12; Agardh, Spec. Gen. Filic. 47.

Chatham-Island. Capt. Anderson.

Mr. Travers also refers to this fern in his journal, unless the notes apply to P. scaberula. It is widely and often gregariously dispersed through extratropical Australia.

PTERIS SCABERULA.

Ach. Richard, Voy. de l'Astrol. i. 82, t. 11; J. Hook. Flor. Nov. Zeel. ii. 25; Hook. Spec. Filic. ii. 174, tab. 93, A; P. microphylla, A. Cunn. in Hook. Comp. to the Bot. Mag. ii. 366; Allosorus scaberulus, Presl, Tentam. Pteridograph.

Common on open places of Chatham-Island.

The diagnostic limits of this species require a still fuller elucidation. The pinnules of the Chatham-plant are in general more deeply divided than those of the plant illustrated by Ach. Richard, but precisely analogous forms occur in New Zealand. It seems by no means improbable, that this fern through middle forms is connected with Pt. aquilina, quite as great aberrations being known to exist in Asplenium marinum.

PTERIS INCISA.

Thunberg, Prodrom. Plant. Capens. 171; Flor. Capens. 733; Schlechtend. Adumbrat. Fil. Cap. 44, t. 25; Blume, Filic. Jav. 212; Agardh, Spec. Pterid. 76; Rawson & Pappe, Syn. Filic. Afric. Austr. 26; Hook. Spec. Filic. ii. 230; J. Hook. Flor. Tasman. ii. 140; P. vespertilionis, Labill. Nov. Holl. Plant. Specim. ii. 96, t. 245; R. Rrown, Prodr. Flor. Nov. Holl. 154; J. Hook. Flor. Antarctic. i. 110; Fl. Nov. Zeel. ii. 26; P. Australasica, Desv. Prodr. 302; P. glaucescens, Bory in Willd. Spec. Plant. v. 396; P. cruciata, Kaulfuss in Sieber, Synops. Filic. 79; P. Brunoniana, Endl. Prodr. Flor. Insul. Norfolk. 12; P. montana, Colenso in Tasm. Phil. Journ. accord. to J. Hook. Flor. Tasm. ii. 140; Lithobrochia incisa, Presl, Tentam. Pteridogr.

Chatham-Island. Capt. Anderson.

In Continental Australia this fern ranges from Mount Gambier through moist forests to New South Wales. Its precise relation to several allied plants is not yet fully established.

ADDITIONS.

ARALIACEÆ.
HEDERA CRASSIFOLIA.

A. Gray, Botany of Wilkes's Unit. Stat. Explor. Exped. 718; Aralia crassifolia, Banks & Soland. accord. to All. Cunn. in Annals of Nat. Hist. ii. 214; Walp. Repertor. Bot. Syst. ii. 430; Hook. Icon. Plant. t. 583; J. Hook. Flor. Nov. Zeel. i. 96.

Chatham-Island. Dr. Dieffenbach.

A fragment of this plant appears also to be extant in Mr. Travers's collection.

This and the two following plants have been introduced into this work on the authority of Dr. J. Hooker.

RUBIACEÆ.
COPROSMA PROPINQUA.

All. Cunn. in Annals of Nat. Hist. ii. 206; Walpers, Repert. Botan. Syst. ii. 463; J. Hook. Flor. Nov. Zeel. i. 109.

Chatham-Island. Dr. Dieffenbach.

Probably one of the species found by Mr. Travers is referable to this plant.

VERBENACEÆ.
AVICENNIA OFFICINALIS.

Linné, Spec. Plant. edit. prim. 110 (1753); J. C. Schauer in Cand. Prodr. xi. 700; A. resinifera, Forst. Prodr. 246; Plant. Esculent. 72; A. Rich. Voy. de l'Astrolabe, 195; Decaisne, Herb. Timor. 74; A. tomentosa, R. Br. Prodr. 518; Wallich. Plant. Asiat. Rarior. iii. 44, t. 271; Miquel in Lehm. Plant. Preiss. i. 353; Wight, Icon. Plant. Ind. Orient. 1481; J. Hook. Flor. Nov. Zeel. i. 204.

Chatham-Island. Dr. Dieffenbach.

The plants constituting the genus Avicennia are deserving of a new critical exposition, but for which the material in our collection proves insufficient. Apparently only one species exists in Asia and Australia. In Continental Australia it fringes widely many tracts of the muddy coast and æstuaries subject to the regular flow of the tides, producing more frequently a narrow-leaved variety on the tropical shores and a broad-leaved form in the extratropical latitudes, the former variety having sometimes the leaves narrow-lanceolate and long-acuminate, the length exceeding many times the width. It occurs as far south as Wilson's Promontory, and will therefore probably yet be met with on the north-coast of Tasmania.

Jacquin's A. tomentosa of the western hemisphere, Selectar. Stirpium Americ. Histor. 178, t. 112, fig. 2 (1763); Linné, Gen. Plant. edit. Vienn. 579 (1767); Murray, Syst. Veget. 484 (1774); Humboldt, Bonpl. & Kunth, Nov. Gen. et Spec. Plant. ii. 283, is regarded by Schauer as distinct from Linné's A. officinalis. Blume (Bijdragen tot de Flora van Nederlandsk Indie, p. 821), Wight (Icon. Plant. Ind. Orient. iv. num. et tab. 1481 & 1482) and Miquel (Flor. Ind. Batav. ii. 912), distinguish a second Asiatic species as A. alba, whilst R. Brown unites the West African A. Africana (Beauvois, Flore d'Oware, i. 79, t. 47) and Walpers (Repert. Bot. Syst. iv. 131) in addition to this the genuine American A. tomentosa with our plant. In Schreber's edition of Linné's Materia Medica, 158, the Asiatic and American plants are also combined as A. tomentosa. Lamark's principal figure (Encycl. Méthodiq. t. 540), quoted by Walpers, is evidently not referable to our plant. The figures in the works of Rheede (Hort. Malabar. iv. t. 45) and of Rumpf (Herbar. Amboinens. iii. t. 76) are sufficiently expressive of the Australian species.

Linné (Flor. Zeilan. 23), Willdenow (Spec. Plant. iii. 395) and also Schauer and Walpers have offered respectively a detailed synonymy of this and congeneric plants.

Eurybia Traversii in a flowerless state bears considerable resemblance to Avicennia officinalis.

EXPLANATION OF THE MAGNIFIED ANALYTIC
FIGURES OF THE PLATES.

PLATE I.—*Gingidium Dieffenbachii.*—1, male flower; 2, front view of a stamen; 3, back view of a stamen; 4, side view of a stamen; 5, side view and 6, front view of a male flower after the lapse of the stamens.

PLATE II.—*Eurybia Traversii.*—1, unexpanded capitulum; 2, expanded capitulum; 3, a complete bisexual flower; 4, corolla and genitalia of the latter; 5, corolla of the same laid open; 6, stamina; 7, pollen-grains; 8, style of bisexual flower; 9, unisexual flower complete; 10, its corolla and style; 11, its style separate; 12, bristles of the pappus; 13, alveoles of the receptacle; 14, achenium; 15, its longitudinal and 16, its transverse section; 17, embryo.

PLATE III.—*Senecio Huntii.*—1, indumentum; 2, unexpanded capitulum; 3, expanded capitulum; 4, bisexual flower complete; 5, its corolla and genitalia; 6, its corolla cut open; 7, stamens; 8, pollen-grains; 9, style of bisexual flower; 10, unisexual flower complete; 11, back view of its corolla; 12, involucre and receptacle; 13, achenium; 14, the same longitudinally dissected; 15, embryo; 16, transverse section of achenium; 17, bristles of pappus.

PLATE IV.—*Senecio radiolatus.*—1, unexpanded capitulum; 2, expanded capitulum; 3, longitudinal section of capitulum; 4, bisexual flower complete; 5, its corolla separate; 6, corolla laid longitudinally open; 7, stamens; 8, pollen-grains; 9, style of bisexual flower; 10, unisexual flower complete; 11, dorsal view of the same; 12, its style; 13, involucre and receptacle; 14, alveoles of the latter; 15, achenium dry; 16, achenium moistened; 17, longitudinal and 18, transverse section of achenium; 19, embryo; 20, bristles of pappus.

PLATE V.—*Leptinella Featherstonii.*—1, unexpanded capitulum; 2, expanded capitulum, with part of the uni- and bisexual flowers removed; 3, a bisexual flower complete; 4, the same laid open; 5, stamens; 6, style of the bisexual flower; 7, pollen-grains; 8, a complete unisexual fertile flower; 9, longitudinal section of the same; 10, achenium cut transversely; 11, side views of embryo.

PLATE VI.—*Leptinella potentillina.*—1, unexpanded capitulum; 2, expanded capitulum with part of the bi- and unisexual flowers removed; 3, involucre and receptacle; 4, a bisexual flower; 5, the same laid open longitudinally; 6, stamens; 7, pollen-grains; 8, style; 9, unisexual flower; 10, the same longitudinally cleft; 11, transverse section of achenium; 12, side views of embryo.

PLATE VII.—*Myrsine Chathamica.*—1, fruit seen from below ; 2, fruit seen from above ; 3, putamen ; 4, longitudinal and 5, transverse section of fruit ; 6, embryo.

INDEX.

L

By Authority: JOHN FERRES, Government Printer, Melbourne.

Gingidium Dieffenbachii, *FM*

Eurybia Traversii *FM*

Eurybia Traversii *FM*

S. Schonfield del. et lith.   F. Mueller direxit   De Gruchy & Leigh imp

Tab. V

Leptinella featherstonii.

Leptinella potentillina. *FM*

Myrsine Chathamica *F.M.*